The Art of
Creative Teaching:
Primary Science

The Art of
Creative Teaching:
Primary Science

Big Ideas, Simple Rules

Alan Haigh

Longman
is an imprint of

Harlow, England • London • New York • Boston • San Francisco • Toronto
Sydney • Tokyo • Singapore • Hong Kong • Seoul • Taipei • New Delhi
Cape Town • Madrid • Mexico City • Amsterdam • Munich • Paris • Milan

Pearson Education Limited

Edinburgh Gate
Harlow CM20 2JE
Tel: +44 (0)1279 623623
Fax: +44 (0)1279 431059
Website: www.pearsoned.co.uk

First published in Great Britain in 2010

ISBN: 978-1-4082-2802-9

British Library Cataloguing-in-Publication Data
A catalogue record for this book is available from the British Library

Library of Congress Cataloging-in-Publication Data
Haigh, Alan.
   The art of creative teaching: primary science : big ideas, simple rules /
Alan Haigh.
      p. cm.
   Includes bibliographical references.
   ISBN 978-1-4082-2802-9 (pbk.)
   1. Science–Study and teaching (Elementary) 2. Inquiry-based learning. 3.
Effective teaching. I. Title.
   LB1585.H25 2010
   372.3'5044–dc22
                              2009031188

10 9 8 7 6 5 4 3 2 1
13 12 11 10 09

Text design by Sue Lamble
Typeset in 9pt ITC Stone Serif by 3
Printed in Great Britain by Henry Ling Ltd at the Dorset Press, Dorchester,
Dorset

The publisher's policy is to use paper manufactured from sustainable forests.

Contents

Prologue

In these days of the internet and sophisticated search engines there is an even greater need for a book which has all the answers in one place; a coherent overview as opposed to an atomistic episodic picture. This book tries to provide teachers and beginning teachers with the skills and knowledge needed to teach Primary science in a **creative** way that will stimulate and motivate children and teach them to become independent thinkers and learners.

This book is a follow-up to *The Art of Teaching: Big Ideas, Simple Rules*. It follows the same reductionist philosophy and the Keep It Simple (KIS) approach wherever possible. It is an attempt to meet the need of teachers not only to improve their teaching of Primary science but also to enjoy their teaching of Primary science.

We all know that in the world of SATs, Primary science usually fares well compared with English and Maths. However, in its report *Success in Science* (2008), OfSTED states that it is unhappy about the teaching of science at both Primary and Secondary level. In fairness, this 2004–2007 survey does acknowledge that the pedagogy of science teaching has been driven by the testing and assessment regime and that the results are satisfactory. However, it points out that teaching in many cases is not satisfactory; on the contrary it is weak. There is an over-reliance on published materials and restrictive methodologies and the science knowledge and understanding of the teachers needs further professional development. (This in my opinion is one factor that has led to a deskilling of teachers in aspects of planning and a developing culture of reliance on pre-prepared lessons.) Interestingly, a key finding of the survey is that **'the schools ... with the highest or most rapidly improving**

standards ensured that scientific enquiry was at the core of their work in science' (p. 5).

Was the high achievement the result of scientific enquiry or was the scientific enquiry the result of high achievement? Put another way, did the enquiry approach produce the best results, or, because the schools were already good at science, had they the time and inclination to approach the subject with an enquiry-focused methodology? A sort of chicken and egg situation?

From my own experience I would say that these two propositions are not mutually exclusive. On the one hand, the work of Shayer and Adey in the 1980s on Cognitive Acceleration through Science Education (CASE) demonstrated the impact of the 'Hands on; Brains on' approach to the teaching of science which resulted in the improved attainment at GCSE of science as well as English and Maths (Shayer and Adey 2006). Interestingly, the Specialist Schools and Academies Trust (SSAT) is running a pilot project in clusters of primary schools, one aspect of which is to look at the improvement in science pedagogy and its impact not only on science but also on Maths and English at KS2.

From my own work with teachers and student teachers, when they gain the confidence to approach their science in this way they improve not only their science teaching but also their general pedagogy and children's learning.

The two propositions below are almost self-evident but worth stating:

- The enquiry process is at the heart of Primary science (as I would also argue it is at the heart of every other subject).
- Science and reasoning go hand in hand in Primary science (as I would also argue that they do for every other subject).

If we accept these two propositions then **the teaching of Primary science through enquiry and reasoning will benefit the whole of your teaching and the children's learning**.

Whether you consult the curriculum reviews of Sir Jim Rose or Professor Robin Alexander (both due out in the summer of 2009), or the *National Curriculum Handbook* (DfEE 1999), you will find that, tucked away behind the various curriculum descriptors, there is a reference to the key skills of **thinking** and **reasoning**. This intellectual process is fundamental to gaining understanding as opposed to gaining knowledge. It is the key to our aspiration to create children who are the independent thinkers and learners for the twenty-first century. If children are to address this new creative curriculum with an emphasis on enquiry and problem solving which will inspire and motivate them then we need an enquiry-based pedagogy to enable this. Learning to think and reason through Primary science is my answer to this challenge.

Just as with children, if we give teachers the understanding as well as the knowledge they need, then they are more likely to become creative and innovative. This is not disrespecting the electronic and paper publications that are full of pre-planned lessons and ready-made activities, however it is the intent of this book to concentrate on the understanding of the pedagogy and to provide the structure and methods to enable teachers to **'do it for themselves'** in both science lessons and topics.

What a difference this could make! Freedom within boundaries to plan creatively and bring back the enjoyment into teaching and learning. Enjoy!

Alan Haigh
April 2009

Acknowledgements

I should like to thank my wife, Peggy Etchells, for her support and encouragement, particularly when my faith was faltering. I should also like to thank my editor, Catherine Yates, who encouraged me and recognised my passion and commitment to teaching and children's education. I am also grateful to Andrew Machell, the headteacher of Mansel Primary School, Sheffield, who as acted as a further sounding board for me.

Lastly, and from way back in the past, I should like to thank two of my lecturers at Durham University, Dr David Bellamy and Dr Fitzpatrick, under whom I studied two short options, one on Botany and the other on Logic. Both were inspirational teachers and taught me for the first time in my life that learning was enjoyable and not just a chore on the way to passing exams.

Introduction

This book is an attempt to help teachers to rediscover their creative planning skills and give them the knowledge and confidence to plan, teach and assess science for themselves, with less reliance on published materials and the mimicry of activities that they have read or seen on a course. At the heart of this book is an understanding that will allow teachers to 'do it for themselves'!

The book is written in the format of 'Big Ideas' which head each chapter, followed by simple rules and guidance on 'how to do it', which head each of the sections within the chapter.

There are two Big Ideas that spring to mind:

> We need to teach children not only what they need to learn in science, but also how to learn, through science

and

> The good teaching of Primary science is simply good teaching!

This book is based on three key premises:

1   There is far too much science knowledge in the world to know and teach, so just grasp the Big Ideas and leave the detail until later (in other words, Keep It Simple, KIS).

2   It is more important that children think scientifically than that they have a lot of scientific knowledge.

3   Primary science is not laboratory based; it is about the 'world around us'.

nb   The above assertions do not discount the need for children to acquire scientific facts and knowledge.

This book is in four sections:

1 **Planning science:** the 'how' of planning, which will lead to discovery, discussion and dialogue and help children to become independent learners and thinkers. It will also show teachers how to plan more inclusively regarding the wider curriculum.

2 **Teaching and assessing science:** the 'how' of teaching, which will show not only how to get children to intellectually engage in the discovery process and to become independent thinkers and learners but also how to make an assessment **for** and **of** learning.

3 **Science knowledge:** this refers to the teacher's own knowledge and understanding of the fundamentals and principles of Primary science, which are necessary for Key Stages 1 and 2.

4 **QCA and DIY science:** since this is the most common published scheme that teachers use, we need to understand how we can adapt it for our children and our teaching and gain some ownership. The object of this section is to give teachers the skills both to become independent thinkers and planners of Primary science and to make meaningful links to the wider curriculum.

The notion of teachers as independent thinkers and planners and children as independent thinkers and learners is not a new one, but certainly one that could do with reiterating.

nb Much of the book is based on everyday Primary practice and common sense and most teachers will recognise much of what they are doing already. This book is an attempt to share and advance good practice and to give those intuitive teachers a deeper understanding so that they can plan better to do more of what they are good at.

Perhaps more importantly, we are now being encouraged to do just this with Excellence and Enjoyment, the government's

Primary Strategy (DfES 2004), and the recommended liberation of the curriculum and the pedagogy as proposed by Messrs Rose and Alexander.

Teaching with trios

'Trios' emerged as this book was written; it was a way of summarising the teaching strategies and approaches to keep me focused and on track. On reading the first draft it dawned on me that it would make a good guide to the 'big ideas and teaching' philosophy, and as a result I have extracted them and promoted them to the front of the book to help the reader.

Most adults are no different from most children: they remember more easily if information is broken down into chunks. The trio acts not only as an aide-memoire but also as a trigger to unlock the understanding. For example, 'Plan; Hide; Guide' is a recall of lesson planning steps and at the same time triggers the understanding of the pedagogy of **'discovery learning'** which gives children greater 'ownership', and the Piagetian concept of 'cognitive conflict' which stimulates children to 'Stop; Think; Try to make sense' (another trio!). Most readers will recognise these as everyday teaching and learning strategies that you already use to a greater or lesser degree.

This section is an attempt to provide a synopsis of the Big Ideas and Simple Rules that run through the book and to act as a guide to a philosophy and pedagogy of teaching. The list below is intended to help you navigate through the book.

nb The page references will take you to the **'Remember'** which can be found at the end of the section explaining the trio in detail.

1 Plan; Hide; Guide page 6

This is the key planning strategy to discovery learning and a shift to a less passive pedagogy and a more active approach involving 'intelligent dialogue' with children. Plan what you want children to learn but do not tell them explicitly. Next you use questions to direct them down your learning pathway until they 'discover' what **you** want them to find.

2   Keep It Simple

This reiterates the basic philosophy of teaching that underpins this book. Simple does not mean 'stupid'! Simple means simplicity and clarity and it is the key to generating the Big Ideas that underpin the learning and enable the teacher to use a **deductive** approach which takes children from the general concept to the particulars of the subject. In this way a 'framework for learning' is provided for the children to fill in the details and make sense.

3   Teach; Point; Pace

The progression set out above is about having a series of 'teaching points' which will lead the children down **your** 'learning pathway'. This will give the lesson direction and shape focused on the learning objective and hence avoids stalling and maintains pace.

4   Consolidate; Review; Rehearse

This is the process of clarification and consolidation of the learning. It is usually carried out in the plenary, but may also be used as an intermediate strategy and not just at the end. We use this rehearsal to 'make sense' with the children through whole-class feedback and not forgetting to make other links to learning and add a bit of magic where possible.

5   Integration versus Isolation

This section is based on the premise that science makes more sense if it forms natural links with other curriculum areas.

Whether science is the star of the show or has just a walk-on part, does not matter; where feasible it is the preferred Primary method to 'laboratory' science in isolation. Create your own topics or themes or follow the Rose Report (DCSF 2009), as long as the learning is grounded in the National Curriculum Programmes of Study and follows the school's planned curriculum coverage and balance, then all should be well with the headteacher and OfSTED.

6 Engage brain; Speak page 34

Intellectual engagement simply means using our brains, whatever that means! Perhaps it means 'thinking'; however, I would prefer to use the term 'reasoning' which has a less abstract and a more logical feel about it. It is the 'Stop and Think' strategy. We not only have to give children permission to do this but also have to show them how. The results, when children 'engage brain before speaking', can be amazing and illuminating.

7 Content versus Process page 36

We need to recognise that we can consider two aspects of science: (1) the knowledge and (2) the process of doing and thinking scientifically. We also need to realise that delivering the knowledge does not guarantee the thinking process. In my experience children have to be taught and trained to think scientifically.

8 Look; Do; Think page 45

Although 'think' is last in this trio we need to get children to proceed through the enquiry process based on their thinking from the previous steps. Hence:

Look: think what to do based on observation.

Do: carry out your line of enquiry, thinking scientifically how to get the best results.

Think: reflect on what you have found out and try to make sense of it.

9 Dialogue versus Monologue page 50

Much of our teaching is in the form of a monologue, with the teacher doing most of the talking and the children answering questions. This is an appropriate strategy for much of our teaching, particularly at a superficial level; however, there are times when we need to probe deeper into children's understanding and only an 'honest dialogue' will do. This involves both children and teacher 'listening' to each other and making 'intelligent' responses. This is not meant in the IQ sense of the word but in the sense of responding intelligently to what has been said, i.e. making sense of what the other party has said and then making inferences and furthering the conversation based on that.

10 Talk; Reason; Logic page 56

This is probably the most powerful trio of all. If teachers can understand this philosophy and pedagogy then we can truly look forward to teaching and learning appropriate for the twenty-first century.

> Giving children the power of inference and deduction based on logic, and the wisdom of judgement based on validity and reliability, must be one of the greatest gifts that a teacher can bestow on future generations.

11 Logical; Critical; Creative page 68

This trio represents the practical application of trio 10 above. These are the three ways of applying 'thinking', each way having its own intellectual strength similarly underpinned by inference and deduction, but different in the power of their conclusions, from near certainty to 'out of left field'!

12 Cause; Effect; Measurement page 73

This trio represents a basic principle of 'scientific enquiry' and is at the heart of the empirical approach which has been central to the discipline of science since the Dark Ages. The planning and controlling of an investigation to produce reliable results is the object of the exercise. This involves sorting out and controlling the variables so that you can clearly observe 'cause and effect' and then measuring and recording the results. The bonus points for Level 5 attainment come not only for doing this but also for being able to interpret those results based on the evidence and supported by previous science knowledge.

13 Intelligence; Caught?; Taught? page 77

Most teachers recognise that the 'hidden curriculum' needs to be taught more often than caught, especially when it comes to behaviour and school socialisation. How many teachers recognise that intellectual ability is also part of the 'hidden curriculum' that we expect children to absorb as they go along? The argument here is that we can teach this intellectual process to **all** children rather than relying on the cleverer pupils to 'catch on'. Make aspects of this intellectual process your lesson objective sometimes, not just the subject matter.

14 Dialogue; Articulation; Listening page 85

This trio emphasises the shift from a dominating 'monologue' classroom diet to one that includes a balance of 'dialogue', where the children are intellectually engaged by the teacher in intelligent conversations. This takes time and space and children have to be trained to structure and articulate these intelligent conversations.

15 Enquire; Explore; Explain page 89

This is at the heart of scientific investigation, and here is some advice on how to lead children through this process by questioning and without their necessarily realising they are being led. The teacher stimulates the enquiry the teacher wants to happen; encourages the children to explore the routes the teacher knows will be fruitful and then helps the children to explain what the teacher already knew they would find.

16 Know; Do; Understand page 100

This statement was made in the Introduction of the first National Curriculum document and referred to the Programmes of Study (PoS), which are the legal entitlement of the children (DES 1988). It said that the following PoS were what children should know, be able to do and to understand. This, then, was the basis of the teaching of this curriculum, and also the principle of assessing that curriculum. Since then we have moved on in our assessment strategies to emphasise 'formative' assessment as well as 'summative' assessment.

17 Record; Use; Inform page 106

Records serve very little purpose if they are only produced to satisfy your line manager or your tutor. Your record-keeping should inform your practice. This evidence informs practice at two levels: firstly at the level of future lesson planning to meet the class's needs, and secondly at an individual level by identifying children's level of attainment and their needs in order to progress.

18 Simple science concepts page 153

Section 3 explores the scientific principles of Biology, Chemistry and Physics in Key Stages 1 and 2 and spells them out in Big Ideas which are, hopefully, easier to understand then the 'NC speak'. To help there are notes alongside which deal with the science in more detail. The same reductionist philosophy is applied to this section which may mean that at times there is a loss of scientific accuracy. Scientific understanding is the objective for children; scientific accuracy is the objective for scientists, and that will come much later in their education.

19 Creativity; QCA; DIY page 163

Section 4 shows how we can apply our understanding to pre-planned topics or sessions in order to make them our own, while at the same time maintaining the integrity of the science teaching and learning and making links to the wider curriculum (Rose et al.). In this way we can have ownership and use our own creativity to motivate not only the children but also ourselves. The section contains a five-step methodology and worked examples from the QCA Schemes of Work.

Planning science

There are two diametrically opposed views to planning science: the science by numbers, follow me approach, and the laissez-faire, let's see what you can find out approach. Like most things in life, nothing is completely right or completely wrong. In this section I am arguing for a 'Guided Discovery' approach, which means that children do find out but within the teacher's control of the learning. In this way teachers get the learning outcomes they are looking for and the learning is clear to the children.

Lesson planning

chapter 1

Plan, hide and guide

The Big Idea here is that if we want to use an 'enquiry/discovery' approach, then it is important that children's enquiries lead them to discover what we want them to learn, i.e. the correct lesson outcome. The only logical way to achieve this is to hide the outcome in the first place and then lead your children to discover it.

This way of planning may, on the surface, look to be in contrast to the '**share your learning objective with the children**' school of thought. (I have known OfSTED and visiting tutors 'fail' the teacher's lesson for not having this prominently said and displayed.) I seem to be saying '**hide your learning objective**'! That is true, but what I am not saying is, 'Do not give your children an "advance organiser" of the learning they are about to experience'. This is a fundamental principle of constructivist learning theory and D.P. Ausubel.

The danger of sharing the exact objective with the children is that at the end when you ask them what they have learnt, they simply parrot back to you the learning objective, because they know that this will make you happy! However, this does not mean that they have understood and that you have achieved a satisfactory learning outcome.

REMEMBER: PLAN; HIDE; GUIDE

Keep it simple

In the first edition of the National Curriculum (NC) Programmes of Study in 1988 the concept of 'Knowledge, Understanding, Skills and Attitudes' (KUSA) was introduced (DES 1988). This was the analysis of the content by which teachers and OfSTED would assess children's attainments. It also appears to me to be a very useful tool to inform our lesson planning. We need to ask ourselves what is the main objective of this lesson? Is it K or U or S or A, because these generic outcomes require different kinds of lesson plans, from passive 'look, listen and learn' for Knowledge; to active 'guided discovery' with a lot of pupil–teacher dialogue for Understanding. This initial analysis enables the teacher to decide what kind of lesson planning and organisation will be required. (One size does not fit all!)

This book is dealing with the teaching of science and in particular the skill of planning and teaching through an enquiry/discovery approach. Not all science is taught in this way, some lessons are knowledge based and have to be learnt, others are skill based and have to be practised. Hence, throughout Section 1a we are looking at the planning of lessons where **understanding** a concept or an idea is what we want to achieve and rather than telling or showing, we want the children to **'discover for themselves'**.

The first step is to arrive at a simple lesson objective by looking for the Big Idea behind the science. If we look at the 'E' series in the QCA Schemes of Work, e.g. 1E 'Pushes and Pulls' to 6E 'Forces in Action', then the Big Idea behind movement is that nothing moves unless a force makes it do so and if something stands still it is because the forces acting on it balance each other out. If you want children to look at the forces involved in floating and sinking, because it is not only fascinating and part of understanding the world, but also one of the most common experiences they had in their early years of schooling, then **drill down** logically from the Big Science Idea to the **simplest science idea**.

REMEMBER: KEEP IT SIMPLE

Search for the simplest Big Science Idea

If we take the example of sinking and floating we know it is about density and that density is something about weight and size. The volume of water displaced by the object depends on these two factors and we probably know that if the volume displaced is enough the object will float, if not it will sink. Instead of beginning with ideas of density or displacement or 'Weighing in Air and Water' (section 4 of Unit 6E 'Forces in Action', QCA Schemes of Work), **why not start with the Big Science Idea that not all big things sink and not all little things float?**

This is often easily learnt by very young children through play. The next logical step after 'drilling down' to the bottom line is to look for the scientific progression which will lead you to the Big Science Idea of density.

nb The lesson objectives you choose will be the **simplest Big Science Idea** and, as we have said, before we start to plan our lessons decide whether they are K or U or S or A.

We now proceed to the next step, which is to look for the scientific progression of the Big Ideas **that are appropriate to the age, stage and experience of your children.**

Look for the scientific progression of the Big Science Ideas

In this case, we might proceed as follows:

1   We start with 'some things float and some things sink'.

2   We engineer results that show that 'not all big things sink and not all little things float'.

3   We juxtapose weight and size by using some large relatively light things and some small relatively heavy things, thus addressing the Big Idea that it has something to do with 'how heavy it is for its size' that determines whether an object sinks or floats, e.g. a large fat potato that surprisingly floats and a small piece of teak that surprisingly sinks. (Make sure that you have tried them out first!)

4   The next Big Science Idea we approach is the 'volume of water that is moved out of the way when an object floats' (displacement). You simply watch the water level rise. Eureka! You can measure it if you think that the Y6s are ready for a more accurate scientific approach.

5   Finally you can **talk about** the concept of the force of **gravity** pushing down and the **upthrust force** from the displaced water pushing up. If the upthrust force can equal the gravity pushing down then the **forces are balanced** and the object will be stationary and will float, if not it will move downwards and sink! (The Plimsoll line on a ship measures that displacement.)

nb   If any bright spark mentions density then you can simply say that it is the ratio of weight pushing down to volume of water pushed out of the way causing the upthrust. Hence it is **mass/volume** and they'll learn all about that in secondary school and realise that if 'mass = volume', i.e. $1/1 = 1$, then the object will float. (It will float higher in the water if the ratio is less than 1.) On the other hand, if the volume displaced is not enough to support the weight, e.g. 2/1 = 2, then the object will be twice as dense as the water and will sink!

nb When I first went to grammar school I was taught 'mass/volume = density' and did not understand a thing! 'Put the numbers in, Haigh, and don't be such a pain asking why it works!'

As I have said earlier, this suggested five-step sequence was not designed to be taught in one sequence all together, but just the steps appropriate to the level of the key stage required.

✳ eg If you were teaching a Y6 class and you had assessed that they understood very little of sinking and floating and the forces operating, then it may be useful to **talk them through steps 1 and 2** followed by setting them an enquiry lesson which involved sinkers and floaters which produced **cognitive conflict** (i.e. surprising results), as the theorists would say, e.g. potato and teak referred to above.

This mini-investigation could lead you to the next lesson of setting an investigation to see what happens to the water level using a Eureka can and measuring volume displaced and mass and seeing what the Y6 made of the results.

If you really wanted to set them a problem to test their understanding of the concept of floating and sinking then challenge them to make a 50 g ball of plasticine change from a sinker to a floater, but no trial and error. They have to work out the dimensions of their boat first and then test it!

Give the children an advance organiser

In simple terms this means give the children an idea of where they are going so thcy can make sense of their journey, or, as someone once said to me, 'Don't lead children down dark tunnels.' Good teachers have always been clear about what they wanted their children to learn and showed them how to go about it, and resourced the lesson appropriately. With the advent of OfSTED inspections this became a mantra of displaying the lesson objective, often in incomprehensible NC speak!

This really takes us back to the dilemma of sharing our lesson objective with the children mentioned at the beginning of this chapter. I don't think we can apply this mantra unthinkingly and in the same way for every lesson! In discovery lessons we do need to hide the learning outcome and make the lesson objective child friendly and give a much more generic '**advance organiser**', otherwise the children will discover what you have told them they will!

✱ eg   Take the Big Science Idea number 3, above:

'Floating depends on how heavy an object is for its size.'

You may want to investigate this with your Y3s, and then it might be more appropriate to say something like:

'We are going to look at some big things and some small things and see which make the best floaters.'

Hide the real learning outcome and guide the children to it

Now we are clear what exactly we want our children to discover, in this case let's say the Y3s are going to find out that it is about being **'heavy for your size or being light for your size'** that makes the difference, then you need to engineer the investigation so they stumble across this phenomenon. In other words, **the learning objective emerges as the learning outcome!**

As we have argued above and contrary to popular belief, we do not tell the children the learning objective as such but allow them to discover it. We **hide** it and **guide** them to it through questions and suggestions. This will be covered in detail in Section 2 'Teaching and assessing science'.

This approach encourages children to raise questions and seek explanations (key learning skills that we need children to practise), and encourages intelligent dialogue with the teacher. (Later on there will be more about teaching children to behave intelligently and become independent thinkers and learners.)

If we want to avoid a laissez-faire, 'do what you want' approach by the children, then we need to know the **learning pathway** down which we wish them to proceed. The key to the learning pathway is to have milestones along the way. These milestones are known as the **Key Teaching Points**; following these enables you to provide **pace** and reach the lesson outcome without stalling or sprinting.

Learning pathway and Key Teaching Points:

1 Some things float and some things sink.

2 Things that sink are usually heavy.

3 There are exceptions to the rule.

4 The exceptions seem to have something to do with weight and size. (Use two identically sized balls, one of rubber, the other of foam besides the potato and teak!) (Notice how we have engineered this by controlling the 'size' variable.)

5 Things that are heavy for their size sink; things that are light for their size float.

Now we know where we are going it is easy to lead them!

REMEMBER: TEACH; POINT; PACE

Consolidate the learning

Most teachers are familiar with the **plenary**; this is a logical end to the lesson where we share the learning with the children. This is good practice; however, it does not mean that we cannot have a plenary during the lesson! We do not have to follow the mantra; we can make intelligent decisions for ourselves and if appropriate we can stop the class at any time and draw the learning together. The point of the plenary, whenever it is, is to **clarify** and **consolidate** the learning for the children.

It is not so much the where and the when but the **quality** of the plenary. It needs to be more than the superficial read/show/tell, although they have their purpose. The teacher is very clear through the planning about the journey and the outcome expected, so rather than ask a vague question such as, 'Can anyone tell me what we have learnt?, guide the children to the learning you are looking for.

∗ eg   'I think we have found out some surprising things about floating and sinking this lesson. Can someone tell me what surprised them?'

In this innocent way you have put the children on the track you want them to take and started a **dialogue** which you can then direct them to and emphasise the five Key Teaching Points, which is the learning outcome you are trying to achieve.

Once we have clarified and consolidated we need to finish by making **links to extend the learning**.

Firstly, we need to **ground** the learning in the everyday world wherever possible. In this case, examples are ships floating high on empty so taking on roof tiles as ballast in the sixteenth century when returning from the Low Countries; Plimsoll line, etc.

Next we need to make links to the next science lesson if appropriate:

> 'Now we understand something about sinking and floating we are going to look at working out how much load we can safely carry without sinking our boat.'

Wherever possible add the wow factor: **'Wonderment and magic!'**

There is plenty of it in the world of science, but often not enough is made of it.

Try wherever possible to finish with a

> 'Did you know ...?'

This can be a fact from recent news, science or the **Guinness Book of Records**, or something amazing you know. The sinking of the **Titanic** provides some good knowledge about heavy things floating and what happens when ships take on too much ballast, in this case water!

REMEMBER: CONSOLIDATE; REVIEW; REHEARSE

Use a lesson planning methodology

If we were to have a plenary on this chapter we should be consolidating the **methodology for Plan, Hide and Guide lessons** that we have developed:

1 KIS and locate the NC source

2 Simplest Big Science Idea

3 Scientific progression of Big Science Ideas

4 Advance organiser

5 Hide the learning outcome

6 Construct a learning pathway to guide

7 Consolidation and wonderment

nb 'Floating and sinking' was deliberately chosen as an illustration because in my experience it is a challenging concept for children and many teachers, and Physics, in my opinion, is the most challenging of the three core sciences.

Below is another example taken from Sc4, without the explanations, in order to show that it is **not a complicated and time-consuming way of planning.**

Light and shadows

Plan

1 Light travels from a source in straight lines and causes shadows (NC Programme of Study Focus; KS2 Sc4 3a, 3b).

▶

2   Simplest Big Science Idea:

Without light there is darkness.

3   Scientific progression of Big Science Ideas:

— Some things make their own light, others reflect the light that shines on them.

— When light shines on some materials it is absorbed and does not reflect or pass through. This causes a shadow to be created behind where the light could not pass.

— Shadows change as the light source moves.

4   Advance organiser, e.g.

'**We are going to investigate light and the shadows that it casts.**'

5   Be clear about the specific learning objective you wish to see as an outcome so that you can hide it and allow it to emerge through the investigation, e.g.

'**Light comes from a source and depending on where that source is the shadow will alter.**'

6   Create a learning pathway using Key Teaching Points:

1   There are different sources of light. (Sun; electrical and battery; fire)

2   Shadows are caused when light cannot pass through an object. (Opaque)

3   Shadows change when the light moves or the object moves.

Points 1 and 2 can be taught by direct whole-class teaching methods.

Point 3 is best taught by allowing the children to investigate and discover for themselves.

7   Consolidate in the plenary and introduce wonderment such as eclipses or the fact that light from the Sun travels at a speed of 186,000 miles per second, and that's faster than Superman!

Planning science topics and themes

chapter 2

Science learning is more effective when not taught in isolation

The Big Idea here is that if we want children to make sense of science then it needs, wherever possible, to be **naturally** linked to their learning in other areas of the curriculum and, where appropriate, relate to the everyday world. This is not out of line with current thinking on Primary cross-curricular design promoted in the Rose Review (Rose 2009).

This is not as difficult as you might think because, as has been said, 'science teaching is just good teaching'. The **enquiry approach** and the **Plan, Hide and Guide** are at the heart of topic and theme work and the same methodology can be applied to planning subjects other than science.

> The grounding of the learning in the everyday world and making links to other subjects where appropriate is very effective in bringing science out of isolation.

In Section 4 we will look in detail at planning at the level of Topic, as opposed to Lesson. Unsurprisingly, you will find it is not much different. We still need to look for the Big Science Ideas and organise their progression, which will then feed down to lesson-level planning.

There are essentially two different approaches to planning at this level:

1 Science drives the **topic** and reaches out to other subjects.
2 Another **theme** or subject drives the **topic** and reaches out to science.

REMEMBER: INTEGRATION VERSUS ISOLATION

Use science to drive the topic

If we take the science illustrated in Chapter 1, e.g. floating and sinking, then this could have been part of a science-focused topic such as:

'Movement on land, water and in the air'.

This would cover the National Curriculum Sc4 'Physical Processes', section 2 'Forces and Motion', and relate to QCA Units 1E, 2E, 4E and 6E.

nb This approach means you are firmly grounded in the Programme of Study and can use any of the QCA guidelines as you see fit and not become a 'QCA slave' where you follow a programme as laid down for x hours over y weeks. More of this in Section 4 'QCA and DIY science'.

Below is a **topic planning methodology**:

1 KIS

Choose the biggest science idea which will give you and the children an overview, but make sure that it is in alignment with the appropriate part of the Programme of Study. In this case it could be:

'When something moves or stops moving, be it on land, water or in the air, then something makes it! This something is called a "Force".'

2 Science progression

Again the choice is the teacher's as to what you wish to include, and this will depend on the age and ability of the children and the time available. In this case it could be:

▶

- Forces that make things move and stop on the land (e.g. pushing, pulling, turning, slipping, sliding, braking).
- Forces that make things move and stop on the water (floating (upthrust) and sinking; wind power; water resistance).
- Forces that make things move and stop in the air (flying (upthrust); falling; air resistance).

3 **Major science concepts**

Use the major science concepts that are behind the Big Science Idea of the topic. This step will give the science in the topic **coherence**. The concepts that seem to be running through are:

- Force
- Gravity
- Friction
- Balanced forces
- Air resistance
- Newtons
- Weight

The most important thing here for the teacher is that they have a **short list** of **key science concepts** which will guide their teaching though the topic and achieve the **scientific learning outcomes** they want for their children.

4 Decide the number and sequence of science lessons which will form the learning pathway through the topic, e.g.

Movement on land; movement on water; movement in the air.

Then break these down into key science lessons, e.g.

Lesson 1: Pushing; pulling; sliding and stopping (the concepts of force and friction).

5 Move on to the 'Plan, Hide and Guide' methodology for individual lesson plans.

6   Now we have planned the **'science spine'** of the topic we
    can **reach out** at the appropriate times to make
    connections with the wider curriculum, e.g.

    When doing floating and sinking as illustrated above,
    the topic would naturally connect with History
    (invaders to the Second World War); Geography
    (oceans, rivers, floods, coastal erosion); Art
    (seascapes); Music (inspired by rivers and seas, e.g.
    Smetana's 'Vltava' or Britten's Sea Interludes).

    The connections could be endless; however, we have to
    make sure of their place in the NC and the school's
    medium- and long-term plans.

What a difference this could make! Freedom within boundaries
to plan creatively and bring back the enjoyment into teaching
and learning. Enjoy!

nb   For detailed examples see Section 4 'QCA and DIY science'.

Plan a topic and reach out to the science

Other subject-driven topics

The same six-step methodology as above could be used for any other subject that you have chosen to drive your topic, simply substitute your own subject wherever the word 'science' is used. The planning principles are the same. Step 6 simply changes to making connections to science that are appropriate to the children and the medium- and long-term plans of the school.

> ***eg** If it had been a **Geography-**driven topic based on a theme of 'Water and its effects on landscapes and people' (Geography KS2 6, Breadth of Study c), and entitled 'The Floods', then natural links could be made to the **force** of water and **floating and sinking**, to reach out and fulfil the science aspect of the topic.

Thematic topics

Sometimes this integrated way of planning is focused not on any particular subject, but on a **theme**. At the time of writing, with credit crunches and the like, an appropriate theme might be 'Money'.

In thematic planning it is the **Big Ideas of the theme** that drive the planning. You then reach out from those ideas to the subject content appropriate for the age and stage of your children and in line with the school's medium- and long-term plans.

* eg Money links obviously to Maths, History, Geography (buying and selling currencies), and to Personal, Social and Moral education with respect to spending and poverty. Whether you give each aspect equal time or not is up to you. You may feel strongly about indebtedness and poverty, and put more emphasis there.

nb There is no link to science here because it would seem contrived, e.g. magnetic and non-magnetic coins?

The above example was chosen to illustrate that you do not have to have complete curriculum cover when planning a topic, as long as in the medium and long term you provide a **'broad and balanced curriculum which fulfils the NC PoS'**. For other examples see Section 4 'QCA and DIY science'.

The beauty is that you have some freedom of choice and can take ownership of the curriculum in a small and disciplined way which will give both you and the children pleasure.

Teaching and assessing science

This section is about moving the pedagogy from the prescriptive and didactic approaches identified by the OfSTED *Success in Science* survey in 2008, to one of more **enquiry and dialogue** with the children. Much of the science that is taught in our schools is done 'to the children', rather than 'with the children'. This is understandable in our measurement-driven curriculum, and it is not surprising that many teachers feel insecure about teaching science and that science usually has bad press: 'it hurts your brain', rather than 'it's great fun and exciting'.

It is important to **start where you are at**; beginning in a prescriptive way is an acceptable start. In this section you will be overloaded with advice from: Look, Do, Think; Talk, Reason, Logic; Enquire, Explore, Explain, to teaching children to 'Think straight'. Select the advice that you are comfortable with and will suit your children, and put it into action one step at a time! In a word, make it **manageable**; do not set yourself up to fail!

The object of this section is to move you on to a more adventurous approach to teaching Primary science in particular and other subjects in general. There is no doubt that many of you will recognise things that you already do, as it is simply good practice. What we want to achieve is for you to do more of that and to add to it approaches that you do not use at present but you feel motivated to try.

Common sense tells us that when we begin, we begin on the learner slopes before we move up the mountain, or else we could find ourselves in trouble. The quantum leap for many of our children is to move from a dependent and instruction-driven mode to real intellectual engagement. Hence the adage

'Hands on; brains on'

(which I pinched from someone about 25 years ago!).

nb The 'trios', as explained at the beginning of the book, are meant to be a simple guide and aide-memoire, but remember you cannot do them all at once. Look at your teaching and your class and begin with what you think you need to do first. Many of the things you will already be using in one form or another; decide on the first step, take one step at a time and do it well!

Principles of learning science

This philosophy of learning science is based on two principles:

1   Teach from the **'Big Ideas' of science**: This section explores the principles of scientific enquiry (Sc1), and the next section looks at the Big Ideas behind Biology, Chemistry and Physics.
2   Teach according to the **'deductive principle'**: Start with the simplest overview (Big Idea) and progress down to the detail in ever decreasing circles.

The cognitive psychology of learning science is based on the following principles:

- It is about children making sense for themselves with the teacher's help.
- This psychology is generally recognised as Piagetian, with later refinements thanks to Vygotsky ('zones of proximal development'; 1978). Further refinement has been made by researchers such as Wood (1986) 'contingent teaching', and

Tharp and Gallimore (1988) who described four stages of progression and emphasised the technique of 'scaffolding'.

■ Margaret Donaldson (1993) described the context which was required for learning: the combination of the abstract with the concrete, i.e. embedding the learning or problem solving in a familiar context. She emphasised the need for children to have an 'inner dialogue' and apply a 'systematic exploration of the mind', which was fundamental to science and maths and probably most other subjects.

■ Ros Driver (1983) reminded us that children do not arrive as empty vessels, but already have previous knowledge and schemata which they have used to build up their own explanations, which she terms 'alternative frameworks'. This reminds us that in our teaching we need to listen to and explore with children their misconceptions and try to dispel them. (Much of science is counter-intuitive, e.g. 'The Sun goes around the Earth because I can see it move across the sky!')

The above principles form a basic overview and general guidance about the teaching and learning of science; the chapters that follow set out the Big Ideas and rules that will provide you with more specific guidance.

chapter 3

Teach children to intellectually engage with the discovery process

Intellectual engagement is about using the brain. However, this can occur at different levels; sometimes it can be quite **'shallow'** and that may be what is required – not all teaching needs to be **'deep and heavy'**. A great deal of learning requires recall and superficial observation and comment, and in that case a lighter touch is all that is required.

Efficiency in teaching and learning is about applying the required amount of 'force to move the object'; to use an analogy from Section 1, and much more 'force' is required to reach the 'deeper levels' of engagement. This is often the case in Primary science teaching and learning if children are to understand and not merely recall, and that is what this chapter is about.

One of my observations of Science SATs at KS2 is that the majority of the questions could be answered by recall. This could explain the 'satisfactory results' but the 'weak teaching' observed by OfSTED. Interestingly, the latest Science SATs are now beginning to include more 'thinking' questions which address that deeper level.

The above is about asking our children to **stop** and **think** before speaking; to think things through and **reason**.

REMEMBER: ENGAGE BRAIN; SPEAK

Teach children process as well as content

A simplification of most subjects is that they can be divided into 'process' and 'content', i.e. the thinking processes and the body of knowledge. We usually recognise the process element as the **'enquiry'** aspect of the curriculum, which requires skills such as problem solving, debate, judgement and reasoning. In this case it is Sc1 enquiry, but this could equally be Ma1, En1, historical enquiry or geographical enquiry, to name but five obvious curriculum areas.

There has been an argument about process and content as long as there has been teaching; from Socrates, whose teachings were almost entirely focused on the process of reasoning and logic, to the Victorian ethos of rote learning and encyclopaedic recall.

If we look at the Programme of Study in Science then Sc1 is all about the process of enquiry and the skills that are required. This does not mean, however, that there is no 'process' requirement in Sc2, 3 and 4. On the contrary, if we look at the aspects of Biology, Chemistry and Physics that require our children to 'understand', then we are demanding much more than recall, we also need them to **'intellectually engage with the idea or concept'**.

It would appear that there are two aspects to intellectual engagement:

1 The process of scientific enquiry (Sc1): an empirical approach to understanding the world based on reliable evidence and the investigative skills to derive that evidence. (Scientists would recognise this as 'scientific method'.)
2 The process of scientific understanding of concepts through fact, evidence and reasoning in Biology, Chemistry and Physics.

Both the above have one thing in common: they rely on a thinking process which I will refer to as **'logical reasoning'**. This is what Edward de Bono (1997) called the 'Fourth R', and in my opinion is somewhat neglected by teachers.

So, how as teachers do we teach this important process? The first step is to recognise the difference between process and content and that when it is necessary to intellectually engage the children at that deeper level, we need to alter our methodology to suit.

`nb` Discovery is not just about practical science investigations; it is also about the discovery of ideas and concepts.

REMEMBER: CONTENT VERSUS PROCESS

Analyse the Sc1 enquiry process

If we look at the Programme of Study for Science, two basic tenets of scientific enquiry are identified for us:

- Link cause and effect.
- Test ideas using evidence.

These are the same two Big Ideas referred to in the above section, i.e. **logical reasoning** and **practical investigative skills**. These are the twin thrusts of Sc1!

There are two things that we can do to give children an understanding of the scientific enquiry process:

1   Give the children a simple child-friendly **overview** of the scientific process at the beginning of the term and keep reinforcing it. 'Science is about':

> Looking
>
> Seeing what is happening
>
> Thinking what it is telling you
>
> Raising questions
>
> Deciding what to do
>
> Planning and doing an investigation
>
> Collecting the findings
>
> Seeing what the findings tell us

2   Give children the opportunity to practise the following skills
    in order to carry out the above:

> Observation
>
> Raising questions based on that observation
>
> Measurement
>
> Fair testing
>
> Controlling variables
>
> Recording evidence
>
> Interpreting evidence
>
> Looking for the science behind the findings

It is important to teach these skills with a separate emphasis,
i.e. looking for that particular lesson outcome.

✳ eg 'Today children we are going to look particularly at several
ways to find out which ball is the best bouncer, but **what we
really want to find out is, which is the fairest way to do it?'**

In this case the teacher will have already provided the
equipment, measuring instrument and a plan of attack for
the children, so they can concentrate on the objective in
question, i.e. discovering how to test fairly.

Similarly, if you want to concentrate on the Recording aspect
of the process, then give the children the experimental results
and/or demonstrate the investigation and capture the results
for them, and then ask 'What do we do with these?' 'How can
we make sense of them?'

Too often children have to struggle through the doing and have
no time or are confused about how to capture the findings and
hence are unable to make sense of them.

Don't teach Sc1 all together

When you look at the above skills analysis it seems obvious that we cannot teach it all at once. The Programme of Study emphasises observation and simple measurement at KS1. At KS2 these skills become more sophisticated, and fair testing and controlling variables are added. There is also the introduction of data gathering and recording to enable children to interpret their findings, the final stage in the sequence.

There would appear to be a natural progression in the above analysis, with the increased sophistication and complexity coming towards the end of KS2. Perhaps the whole-school planning in science should reflect this?

Even though this analysis has led to an 'atomistic approach' to the teaching and learning, there is no reason why teachers cannot begin to 'synthesise' the skills, even at an early age, according to the readiness of the children. (I have seen children in Y3 operating at the same level as some classes in Y6 in their scientific enquiry skills.)

∗ eg When young children become accomplished at observing and drawing what they find, we can move from pictures to diagrams and from teacher script to the children's own narrative. Ways of measuring and testing can also be added in KS1.

It is the teacher's professional decision when to introduce particular skills and then combine them into a **mini-investigation** process. Later you could combine the process into a mini-sequence such as 'Look, Measure and Fair Test', until by Y6 children are capable of **planning and doing their own investigations and capturing and interpreting their findings!**

Teach children to look, do and think

Science was originally based on **imagination** and **theories**, and the ancient Greeks were particularly good at this. For example, several hundred years BC, they imagined that the world was made up of tiny particles, 'electra' in Greek, hence the electron! The scientific era based on an empirical approach did not fully dawn until the seventeenth century. This **empirical** approach, i.e. one based on observation and evidence leading to discovery, is sometimes known as the **scientific method**. It was a way of trying to make sense of the world by **looking, doing and thinking**. (This is the thinking process that we wish to adopt here, but this need not discredit the philosophical and religious approaches.)

Scientific enquiry is a **process of looking, doing and thinking**, with a **beginning, middle and an end**, which I have broken down into four stages.

In order to make sense of this I have overlaid the four stages of enquiry that are in the Programme of Study with the three processes of 'Looking, Doing and Thinking'. Hopefully this will give teachers a better idea of the **learning outcomes** and **Key Teaching Points** they need to emphasise.

Stage 1: Beginning an enquiry

Looking	Doing	Thinking
Observation: ■ Macro: impressionistic ■ Micro: detail	Look and **notice** structure, form, movement, surroundings Look more closely, with a hand lens if necessary	Overview: generalities and commonalities Similarities and differences Patterns and associations Previous appropriate scientific knowledge Inferences: Based on the above, what does it appear to tell us? What questions does it raise? Do you have any proposals for further enquiry?

Stage 2: Planning and investigating

Looking	Doing	Thinking
■ Look for changes; **cause and effect** ■ Look again to make sure	Set up the investigation: ■ Design opportunities to measure cause and effect i.e. See 'what happens if?' ■ Control **variables** ■ Ensure **fair testing** ■ Apply the **appropriate** instrument to measure findings	Decide which ideas to test and how to measure them Consider the constraints and practicalities Consider the possible predictions; decide which to follow and how to **measure** and **record** it Select one question, idea, line of enquiry and its experimental design **Keep an open mind!**

Stage 3: Findings and their recording

Looking	Doing	Thinking
Look for evidence of physical changes, e.g. size, shape, weight, time, colour, reaction etc.	Capture evidence by recording the results:	Concentrate on making sense of what you are seeing
	Drawings	
	Diagrams	Think about knowledge from science and other subjects that may be of help. **Make connections!**
Look at the evidence using appropriate measurement to differentiate the results from each other. (Not to 2 decimal places!)	Tables	
	Charts	
	Graphs	
	Information and communication technology (ICT)	
Look again and repeat to make sure observations are **reliable**.	Narrative, written or spoken	
	etc.	

Stage 4: Interpret and evaluate

Looking	Doing	Thinking
		'Search for meaning'
		■ INTERPRET
		Analyse results and data and look for:
		Cause and effect
		Correlation
		Patterns
		Logical inference
		Extrapolation
		■ EVALUATE
		In your judgement, what is it telling you?
		Did it support earlier predictions? If not, why not?
		Did it suggest any further line of enquiry?
		Could we have made improvements?
		Can we explain it logically?
		Can we explain it scientifically?

I have tried to break down this complex intellectual process into simple steps without 'dumbing it down'. The **biggest mistake** we can make is to assume that we can go through this process (stages 1–4) more or less as a complete sequence with the children, and the children will absorb the method of scientific enquiry.

Use the principle of the teacher doing a stage or more for the children so that they can focus on the particular stage you want them to practise. For example, if you want them to test fairly, control variables and record the results, then give them the 'investigation' part of the exercise.

Ice melting and the effect of different salt solutions

'Here are some ice cubes which are exactly the same size and five different strengths of salt solution. How can you test them fairly to find out which solution dissolves the ice the quickest and record it for me in a table?'

Hence the children can concentrate on setting up the investigation fairly and scientifically, and recording their results.

As said above, we can look at the process one stage at a time: Beginning; Planning and Investigating; Recording Findings; and Interpreting and Evaluating. Once again we can use our creativity to combine these into mini-processes covering more than one stage. In this way we shall synthesise the whole of the enquiry process by the end of KS2.

REMEMBER: LOOK; DO; THINK

chapter 4

Teach children to become independent thinkers and learners

The Big Idea of this chapter is a statement of the obvious. It is also part of the government's Excellence and Enjoyment strategy (DfES 2004) and the ultimate aspiration of all teachers of all ages of children. Probably most teachers achieve this to varying degrees, some by accident and some by design. The object of this chapter is to look at the simple rules that show teachers **how to achieve this aspiration by design**.

Do we rely on it being **'caught'** during the delivery of our subjects or do we specifically **'teach'** these processes? Is it implicit or explicit in our teaching? From my experience it is more the former than the latter. I would argue that if we are going to make this intellectual process **inclusive** for all children and not just the domain of those who are clever enough to make sense of it for themselves, then we need to be far more explicit and teach the strategies!

Allow children to talk

This is the one key factor that needs to permeate the teaching and learning process. It is true to say that teachers do most of the talking in most of the classrooms that I have visited. Children do talk; they answer questions trying desperately to guess what the teacher wants them to say.

They are taught 'Speaking and listening' as part of their literacy and this 'skills-based approach' is helpful, but how does it permeate their everyday thinking and the rest of the curriculum? Apart from the question and answer, listen and tell me what I have just said, which are all one-way from teacher to the child and, I agree, an essential part of the teaching and learning process, when do the children get the opportunity to talk? They often talk when they are not supposed to and get into trouble, but how often do they get the opportunity to reverse the flow down this one-way street?

We need to change this monologue into a **dialogue**, a two-way conversation where children's talk is valued and used to extend their thinking and learning. This **intelligent conversation** between teachers and their children is a much less frequent occurrence than is desirable and I am arguing for a considerable increase in this aspect if we truly want our children to become independent thinkers and learners. If we are to make this culture shift in our classroom then there are certain essentials that need to be fulfilled.

- Create a classroom climate where children feel free to talk to the teacher and their peers and raise questions, issues and dilemmas without fear of criticism or ridicule.

    How is that done? It is not difficult if the teacher is prepared to take the risk. The risk I am talking about is that of children **challenging** the teacher and the system. No, it is not anarchy that is being proposed; children have to learn that much abused slogan, **'with rights come responsibilities'**.

■ Come to an agreement about the **ground rules** of speaking, listening and raising questions. These are common sense and focus on politeness and respect for others and their point of view on the one hand and, on the other hand, needing to have a reasonable argument for raising questions or objections. Children, as we all know, have a strong sense of fairness and gladly accept and contribute to the ground rules process.

■ Give children **time to think**. There are times when quick responses are necessary and children grow up with the culture that kudos is gained by being first to the answer. (Every quiz on the TV is based on this principle.) Children need to know that there are other times when it is more important to **stop and think**; and this is as important as, if not more important than, being first and **you don't have to come up with the right answer**, it's the thinking that counts.

■ They have to appreciate that the **reasoning** is more important than the answer and that **explaining how you got there** is the most important part of the process.

If teachers and children can apply these basic principles there will be a sea change in our classrooms and our children's achievements.

nb Many teachers will recognise some or all of this good practice in their teaching. My proposition is that there should be more of this good practice and it should be deliberately embedded in our teaching philosophy.

REMEMBER: DIALOGUE VERSUS MONOLOGUE

Teach children to reason for themselves

Most teachers do this every day to a greater or lesser degree. Certainly, you will recognise below many of the things that you already do; however, we need to make them more explicit and practise them if we want our children to 'think straight'.

Reasoning is really the opposite of spoon feeding, which is difficult to ignore in our test-driven curriculum. However, it is reasoning, not spoon feeding, which is the key to independent thinking and learning.

This thinking process goes back to Socrates (469–399 BC) and the Greek tradition of questioning and argument, and to Aristotle (384–322 BC) who invented 'rigorous rules of reasoning'. This wisdom from the fourth century BC is the basis of **formal logic** and I would argue that it is also the one intellectual process that should run through not only the whole curriculum, but also through life. In this book the examples centre on the teaching and learning of Primary science, but I am sure that you will appreciate the generic character of this approach.

> **Reasoning**, quite simply put, is a **chain of thought**, otherwise known as an **argument**; however, the chain of thought has to be **valid** and **sound**. This chain of thought involves going from a **premise** to a **conclusion**.

The validity of an argument is about the **truth** and the **soundness** of the **inference**, which is the thinking process that takes you from the premise to the conclusion.

Inference is the logical process of reasoning from the known to the unknown. This inferential process of extrapolating from the

---

known to the unknown, or the past to the future, is commonly recognised as **deduction**.

(This is technically not completely true as logicians differentiate this process into deduction and induction. For this book I am using the one term to cover both processes. For a more detailed look at formal logic try *An Introduction to Formal Logic* by Peter Smith, 2003.)

This excursion into the realms of formal logic is essential if we are going to teach our children to reason based on sound principles and not just make guesses, whether those guesses are logical or illogical.

Why is this so important? As has been said before, **the key to independent thinking and learning is the process of reasoning**, and if this thinking is to be sound then the inferences made by the children need to be logical and valid, to a greater or lesser degree.

Teach children to recognise logic and its vocabulary

Although formal logic has its own rules and language, many of the key words are in our everyday vocabulary but are interpreted a little more precisely in logic. Just as with science vocabulary, children have to be aware of a slightly different understanding of everyday words such as heat, temperature, force, friction, inertia, momentum. All these words are to be found in ordinary conversation, but in science they have a slightly different meaning.

■ The two key words on which the whole of Propositional Logic can be said to be based are:

IF 'x', THEN 'y'

This **conditional clause** is probably the most powerful tool that we have got to drive our inference and lead us to our conclusion.

These two words will further the journey from the known to the unknown; the only thing we have to do then is to test our conclusion for its degree of certainty or, in one word, its **reliability**.

> The IF ... THEN is the first reasoning strategy that children need to learn.

■ The second thing that children need to grasp is that IF is usually followed by a **qualifier**, which by definition alters its meaning. These qualifiers fall into three broad categories:

1   **ALL** (every, everyone etc., and other words that mean the same).

> *** eg**   **If all** dogs have four legs and Deefer is a dog, **then** Deefer has four legs!

This qualifier is what I call **positive** and **certain**. (Well, nothing is certain, is it? Deefer may have had an accident and be a three-legged dog!)

2   **NONE** (No, not one etc., and other words that mean the same).

> *** eg**   **If none** of the members of this class have failed, and Jessica is a member of this class, **then** Jessica has passed.

This qualifier is what I call **negative** and **certain**. (Well, again nothing is certain. Jessica's paper may have been lost by the external marker and Jessica may not have passed!)

3   **SOME** (most, many etc., and other words that mean the same).

> *** eg**   **If some** members of this class have failed, and Josh is a member of this class, **then** Josh may or may not have failed!

This qualifier is what I call **not so certain**, and can be further qualified in a **negative** sense.

> *** eg**   **If some** members of this class have **not** failed, and Josh is a member of this class, **then** Josh may or may not have passed!

■ The third thing that children need to grasp is the

'AND', 'OR' and 'NOT', connectives.

These are three very special propositional connectives, because when permutated they give us very different logical outcomes.

✱ eg 1 Either we are out of petrol **or** the carburettor is blocked.

We are **not** out of petrol.

So the carburettor is blocked!

✱ eg 2 Either Josie is in the classroom **or** in the playground.

Josie is **not** in the classroom.

So Josie is in the playground!

✱ eg 3 Josie cannot be in the classroom **and** in the playground.

Josie is in the classroom.

So Josie is **not** in the playground!

The general formula for the arguments in examples 1 and 2 is:

Either A or B
Not A
So, B

The general formula for the arguments in example 3 is:

Not (A and B)
A
So, Not B

nb Do not worry about the formula; the logic of elimination becomes obvious when you see it in this shorthand.

Logic is a complex intellectual subject at undergraduate level. All that I have tried to do here is give you some simple logical trains of thought to practise with your children.

REMEMBER: TALK; REASON; LOGIC

Practise teaching children to 'think straight'

It is a case of practising the logic and reasoning that permeates our everyday teaching and thinking. We need to identify the logic and then model the thinking and further explain how it works to the children; they soon catch on.

Use children's chains of thought to illustrate the logic of how they came to their conclusion – they may not have appreciated the logic of their train of thought.

Children soon begin to imitate your thinking processes and start to internalise them and think straight for themselves.

Below are four of the logical patterns that you can look out for and practise:

1   **If ... then** exercises:
   — I-Spy variation:

      This is where you are not asking children to guess blindly but you add a clue.

> **✳ eg**   **If** it is in this room and begins and ends with 'w' **then** it might be ...? (window)

      Taking the example of the 'window', the child could come up with the following logic:

      '**If** it is in this room and there are more than one and you can see through them **then** it must be the windows!

— Guess what I am thinking of in 10 questions:

This is where only yes/no answers are allowed. Children soon learn that they can ask only one thing at a time and it can result only in yes or no. At the end of this process they have discovered by elimination and come to a logical conclusion. (In biology the process is known as a 'binary key'.)

✳ eg   Does it have more than 4 legs?

Yes

Does it have 6 legs? etc.

— Number frequently uses the following deductive logic in its reasoning:

✳ eg   'If I have 10 sweets and I give 4 away to my friends, then how many sweets are left?'

✳ eg   'If the answer is 45 and the numbers used are 50 and 5, then what kind of sum must it be?'

"Mathematics is not just a collection of skills, it is a way of thinking. It lies at the core of scientific understanding, and of rational and logical argument."

Dr Colin Sparrow, University of Cambridge, quoted in NC Programme of Study 1999, Mathematics, p. 61

Suduko puzzles are based on these logical principles!

2   **Either ... or** exercises:

> Either A or B
> Not A
> So, B

Or:

> Not (A and B)
> A
> So, Not B

This is the logic of **elimination**. If children realise that there are two things that are **incompatible**, then one of them can be eliminated. How often do we use this pattern with our children? 'You can go to the cinema or the football match!' It is not a situation that they are unfamiliar with in their everyday life, we just need to show them the power of this thought in their learning.

It is one of the logical patterns that underpins logical deduction, I call it the 'alibi principle': things cannot exist in two states or places at the same time.

3   **Multiple choice** exercises:

Not only are these becoming more common in the assessment of children, but they also demonstrate a sound thinking technique:

> One of sequential and logical elimination of options until the choice is reduced and a judgement can be arrived at with more confidence.

> ✳ eg $25 \times 25 = 250$ or 500 or 625 or 2500
>
> Eliminate the 'outliers' (too small, too big), then make a choice based on a rough calculation, e.g. 10 $\times$ 25 = 250, so 500 must be 20 $\times$ 25.
>
> Answer 625.

Is this not a thinking process that can be brought to bear in any situation, in or out of school, **where we have multiple choices and have to weigh up our options**?

4 **Best fit** exercises:

These involve **judgement** when we do not have certainty, only the **'more likely than not'** option. (See next section.)

✳ eg   **Three Little Pigs Alternative Fairy Stories**

Was it more likely than not that the wolf wanted to borrow a cup of sugar rather than to eat the pigs?

Any area of the curriculum which requires children to look at the **evidence** and come up with a **judgement** practises this type of thinking, from Science to History, Geography and Personal, Social and Moral education.

The above four areas enable us to practise logic and to show children how to think straight, and have a great deal of fun at the same time.

Show children how to test their logical deductions

As mentioned in the introduction to Section 2, one of the basic tenets of cognitive psychology is that of children trying to 'make sense for themselves'. Is that not the same as children becoming independent thinkers and learners? The argument just given above is that reasoning is the key to achieving this and, what is more, the reasoning needs to be sound and not flawed if we are going to make sense of it.

The soundness depends on the validity of the deductive inferences that children make. This is what is explored here; **logical connections** and their **soundness**.

Deduction and certainty

Common sense tells us that nothing is certain but some things are more certain than others. We need to show our children that this is true and that they need to **test their thinking** to see how certain their deductions are and that there are not any other possible explanations.

Deductions and evidence

Deductions are based on evidence and this evidence may be obvious or buried as clues. We need to show the children how to look for clues and sometimes to **read between the lines**. At the same time we need to encourage them to **make a judgement** as to how reliable the evidence and clues are.

Deduction and probability

If we look up 'probable' in the dictionary we will find it defined as something like 'likely or credible'. This **criterion of**

probability is another criterion that we need to encourage our children to use to **test their thinking**.

Back to **certainty** again; you can use the analogy of the law which is all based on the probability of being 'proven guilty based on the evidence'.

- In criminal law this degree of probability needs to be almost certain (99%) in order to make a judgement of guilty. (It's the equivalent of our mathematical deductions for instance.)

- In civil law this degree of probability needs to be in the realm of 'it is more likely than not' (51%+) in order to make a judgement of guilty. (This is much more like the everyday deductions that are made in and out of school.)

> We need to show children how to make inferences using deductive reasoning and then checking the security of their conclusions based on probability.

nb All the strategies mentioned above can be simply taught by the teacher modelling the thinking and explaining the logical connections to the children.

Explain answers to children wherever possible to show the logical reasoning! Simply saying 'Correct' often misses wonderful opportunities to teach the reasoning to the rest of the class.

Teach children that there are basically three ways of thinking

In our pursuit of teaching children to become independent thinkers and learners, we have encouraged talk and dialogue; looking for reason and then structuring reasoning logically:

Talk, reason and logic

This directive would seem to sum up the philosophy of teaching children to think so far.

However, once the children have been given these three precursors to independent thinking and learning they need to realise that it is not 'one size fits all'!

There are **different aspects to the thinking process**. These are described in a variety of ways, from the 'Key Aspects of Learning' in the Primary National Strategy 'Excellence and Enjoyment' (DfES 2004), which has five thinking skills and six key skills, to Reuven Feuerstein's 'Instruments' in his Instrumental Enrichment programme (Feuerstein et al. 2004) and Matthew Lipman's 'Philosophy for Children' of the 1980s (Lipman 1988).

In line with my reductionist philosophy and having read much of the published work on thinking, which varies from academically complex treatises to the simplistic 'follow my tips' formula, it has been decided to present the reader with another **trio** to remember:

Logical, critical and creative

These are three ways of thinking which apply 'talk, reason and logic' to a greater or lesser degree and in slightly different ways. These three ways are differentiated by the degree of certainty that results from the thinking process, in other words the confidence that you can have that your answers are correct. We need to illustrate these three ways to our children so they are much more **conscious** of their own thinking processes and can make **independent choices**.

Logical

This term is used here to denote a thinking process which follows the rules of logic and results in an **inference** in which we have complete confidence (the 99% variety we talked about earlier).

* eg **Either** we are out of petrol **or** the carburettor is blocked.

We are **not** out of petrol.

So the carburettor is blocked!

* eg **Either** the salt has disappeared **or** it has dissolved in the water.

The water tastes salty **so** it has not disappeared and **therefore** it must have dissolved in the water.

These are cases where the logic and reasoning leads us to **conclusions** that are **unavoidable**. In this case logic overrules personal judgement.

If we look at the curriculum then in Maths and Science this approach can be particularly appropriate.

Below are some suggestions and questions that may help you to encourage this thinking:

■ Explore and explain the children's answers with them and for the rest of the class.

How did you come to that conclusion?

What are your reasons?

Good thinking!

I see what you mean ...!

How does that follow from that?

How certain can we be that that is absolutely correct?

Critical

This term is used here to denote a thinking process which follows the rules of logic and results in an **inference** in which we **do not** have complete confidence (the 51%+ variety we talked about earlier). This is where we need to decide whether the inference we make is 'more likely than not' and use our personal judgement based on the evidence.

❋ eg   If **some** members of this class have failed, and Alan is a member of this class, **then** Alan may or may not have failed and we will need further evidence.

❋ eg   **If most** of the toy cars with big wheels travel the furthest, **then** the winner **should** be the car with the biggest wheels.

❋ eg   Historical interpretation: Were the Armstrongs and the Elliots, which were two Lowland Border clans, thieves and murderers **or** defenders of the Scottish border?

These are situations which require **critical judgement** and are often signalled by such words as some, not all, either, or, perhaps, possibly – in fact, any words or situations where there is the possibility of **doubt** (hence their clever use in law, politics and the world of commerce).

It is important here for children to seek **further evidence** not only to support their reasoning for one inference, but also to enable them to look at **alternative inferences** and then come to a **judgement of best fit**.

In this case logic **does not** overrule personal judgement based on evidence.

If we look at the curriculum, then in History, Geography, Design and Technology, English and Personal, Social and Moral education, this approach is essential.

Below are some suggestions and questions that may help you to encourage this thinking:

■ Query children's explanation.

   In practice we need to ask children for their reasoning and then encourage them to look for other explanations and other points of view before settling for the answer.

> Can you think why this town was built there?
>
> Could there be another reason?
>
> Which do you think is most likely?

■ Challenge children's assertions by getting them to qualify and quantify where possible.

> 'Big boys are bullies!'
>
> Do you mean that **all** big boys are bullies **or** just **some**?

Children should learn to challenge sweeping statements and assertions themselves in their everyday life. It is the biggest source of **misinformation** that feeds prejudice and fear and not uncommon in the tabloid press.

> ✱ eg 'Teachers claim they are not paid enough!'
>
> 'Armed and dangerous teenagers on our streets!'
>
> 'Immigrant communities swamping Britain!'

Creative

This term is used to denote a thinking process which **does not appear** to follow the rules of logic. Edward de Bono (1997) refers to this as 'logic in reverse', and it sometimes appears that way. In some cases children's brains **leap to the correct answer**; they have no idea how they got there and give the 'It just came to me' type of answer.

This may not be as illogical as it would first appear. It could be that the brain solved the problem by processing the information in a logical way and applying previous experience that was now in the child's subconscious. Such a process is often tacit.

The more overt creative thinking we recognise as **imagination** and **innovation**. All types, be they tacit or overt, come broadly under the term **creativity**.

> ✱ eg Different perspectives, such as those taken by the 'modern' artists throughout the ages, or inventors of new machines or ways of working or theories, or literary giants!

This is sometimes called 'left field' or 'out of the box' thinking by our transatlantic cousins. The important thing here is that we need to signal to children that creativity is where they can use their wildest imagination and turn things on their head.

It gives children that wonderful opportunity to mentally explore **without searching for a right answer.** In fact almost anything they think of can be right.

If we look at the curriculum then in Art, Music, Dance and English, this approach can be particularly appropriate.

Below are some suggestions and questions that may help you to encourage this thinking:

Use your imagination and think of something entirely different from ...

Put yourself in the position of ... (person/animal/ plant/object) ... and look through their eyes. How would you feel/see/act etc.?

Close your eyes and imagine ...

nb It is important to note that these three processes – logical, critical and creative – are not mutually exclusive; in fact they are very much interwoven and they all depend on 'talk, reason and logic'.

Some of the most amazing discoveries in maths and science have been 'inspirational' and not just based on pure logic and experimental evidence. Some of the most amazing music is based on composition capturing mathematical progression.

"Logic will take you from A to B, imagination will take you everywhere."

Albert Einstein

REMEMBER: LOGICAL; CRITICAL; CREATIVE

Look for logic and reasoning in the KS2 Science SATs

One of the facts of life that we have to face is that despite our convictions that we are teaching Primary science in a more enjoyable and intellectually challenging way, we shall still be **judged on our test results.**

Common sense tells us that if we can improve our results by better preparation, then that is a moral obligation in today's curriculum. If you examine the science questions you will see that almost half of them can be answered with very little science knowledge, but a great deal of logical reasoning from the evidence given in the question. Hence it makes sense to practise the logic of the questions and not just rehearse more science knowledge.

Regardless of the proposed ending of KS2 Science SATs that occur in the second week of May and then are externally marked, there still will be a summative assessment of Primary science at some time, to national benchmarks. To this end it could be in our interests to look at how recent science assessments have been designed, since non-externally marked assessment will have to measure the same things. In essence there are three kinds of questions:

1 Those which focus on 'experimental design'.
2 Those which focus on the interpretation of experimental results.
3 Those which rely on a little science knowledge and observation.

nb 1 and 2 are probably the questions that sort the Level 5s out from the rest!

1 Experimental design

Experimental design relies on understanding the principles and reasoning behind the scientific approach, which is essentially empirical and heuristic, i.e. relying on experimental investigation and discovery. For investigations to be scientific, the experimental design has to obey two principles:

- It must show **cause and effect** and measure that relationship.
- The process must be designed so that only **one action** can be directly related to the **corresponding reaction**.

This process is known as 'fair testing' and relies on children grasping the principle of **controlling variables**. All the various experimental conditions must be kept the same and **only one variable** is allowed to be changed (**independent variable**). This **action causes a reaction** which is observed and measured (**dependent variable**).

This **'cause and effect'** produces the experimental results which we now have to interpret.

✳ eg **Q2 Test A 2008 Teeth:** Control variables to give a fair test and vary the kinds of liquid to see cause and effect (i.e. Which liquid is most damaging to teeth?).

✳ eg **Q6 Test A 2008 Bananas:** Control variables to give a fair test and vary the conditions of one to see cause and effect (i.e. The ripeness of the green banana on its own (the control), compared with green banana + ripe apple, after 7 days).

Q7 Test B 2008 Shoes: Only change one factor, control the others to make it fair and measure the effect that results (i.e. Choose either waterproof or grip and keep the test conditions the same but vary the shoes and measure the different effects that result).

2   Interpretation of results

The cause and effect relationship produces the basic logical proposition:

If X then Y

where X is the **action** (independent variable) and Y is the observed **reaction** (dependent variable). Interestingly, on graphs the X is plotted on the *x*-axis and the Y on the *y*-axis.

Many questions rely on children discovering the relationship between the X and the Y or seeing the cause and effect and understanding how it is represented on **block graphs** or **line graphs**. This reasoning then allows them to draw conclusions or make predictions based on the experimental **evidence**.

Sometimes the results X and Y are simply put into a table and children have to see the relationship of the cause to the effect and put in missing values.

nb   Often no previous science knowledge is required, only the powers of deduction based on the results provided.

✳ eg   **Q7 Test A 2008 Floating on salty water:** The saltier the water (cause) the higher the float is (effect).

▶

3 Applying science knowledge

In some cases, applying science knowledge is as simple as recalling the correct scientific term or name; in other cases it is necessary to understand a scientific process or principle and then **apply that knowledge in order to reason**, based on evidence given in the question or the diagram, and produce the correct answer. In fact the knowledge content is not that great and the basic Big Ideas and vocabulary of Biology, Chemistry and Physics are not difficult to acquire.

✱ eg **Q2 Test B 2008 Penguins:** Binary identification keys which are based on the logic 'If A then not B', or 'If not A then B'. Very little science needed, only logical deduction.

✱ eg **Q5 Test B 2008 Flowering plants:** Basic plant anatomy and vocabulary.

✱ eg **Q8 Test B 2008 Parachutes:** Basic physics of forces acting in the air; gravity and air resistance.

✱ eg **Q1 Test A 2008 Solids, liquids and gases:** Basic chemistry of the states of matter.

✱ eg **Q3 Test A 2008 String instruments:** Basic physics of sound relating vibration to pitch.

The above is a rationale not only for doing 'more of the same', i.e. demonstration investigations (whether carried out by the teacher or the children) and rote learning of science facts, but also for teaching the reasoning and thinking underpinning scientific investigation to give the children the independence to enquire and investigate within a disciplined intellectual process.

REMEMBER: CAUSE; EFFECT; MEASUREMENT

Teach children new learning behaviours

This is really the **'Is it caught or is it taught?'** debate. Or the **'hidden learning curriculum'** and the **'Am I allowed to teach the children this explicitly and make this my lesson objective?'**

Do we rely on these processes being **'caught'** during the delivery of our subjects or do we specifically **'teach'** them? Are they implicit or explicit in our teaching? From my experience it is more the former than the latter. I would argue that if we are going to make these intellectual processes inclusive for all children and not just the domain of those who are clever enough to make sense of them for themselves, then we need to be far more explicit and teach children the strategies.

Throughout the curriculum these intellectual processes of reasoning and thinking mentioned above, pervade all subjects: e.g. scientific enquiry; using and applying mathematics; speaking and listening; reading and writing; historical enquiry and interpretation; geographical enquiry and skills; not to mention art, music and design.

> We need to teach these intelligent behaviours to all our children.

Intelligence is the sum of the application of those intellectual processes, and all children can learn these processes to a greater or lesser degree and become more intelligent.

The logic of this argument would lead us to the conclusion that if we taught these processes we would impact on all subjects and show improvement. This would seem to be a more efficient approach to improve teaching and learning than relying solely on specific remediation inputs for each subject. We need to teach the reasoning and thinking and show that it is the same general principle across the curriculum.

In other words, try sometimes to make the learning behaviour your lesson objective, e.g. use of reasoning and logic. In this case we are teaching children to achieve a train of thought which is logically connected step by step, **regardless of the subject matter.**

Besides the reasoning strategies discussed above there are also some more general behaviours we need to encourage in order to **help our children to behave more intelligently**. Below are some suggestions:

- **Impulsivity:** On many occasions we need to get children to **stop and think**! Children need to know that this is a valued expectation of the teacher. We also need to give them a simple method to assess the situation, e.g. 'Clarify the problem/task'. 'Make a plan'. 'Where are we going?' 'Is there anything further we need to know?'

- **Persistence:** Many children give up too easily or don't even begin. This is sometimes because they have poor self-esteem ('I'm rubbish me!'), or they don't know where to begin or continue because the task is too difficult. Tasks need to be challenging and to extend children; however, they need to be accessible to begin with.

- **Carelessness:** This is often as a result of impulsiveness or lack of concentration. We need to get children into the habit of **checking and re-reading** their work whenever possible. We need to give them time and encouragement so they begin to learn that this is an expectation of the teacher. (I accept that there will be times when it will not be appropriate to check their work in our 'hurry along' and 'test-driven' curriculum.)

- **Mind's made up:** This is a natural tendency for children, and adults. We feel much safer with certainty than uncertainty, so it is not surprising that we jump to the first conclusion and stick with it. Unfortunately it encourages a resistance to **flexible thinking** and tends to produce closed minds.

We need to encourage '**open minds**' and discourage 'closed minds'. We need to praise the good start they have made coming to their first speculations and encourage children to consider the 'howevers', the 'buts' and the 'if you look at it another way'.

If the teacher raises these kinds of challenges first then you tend to find that the children will begin to adopt the approach. Further, if we encourage the children to challenge **each other** in this way then, hopefully, they will further **internalise** this behaviour and begin to challenge their own thinking!

■ **Encourage children to ask questions:** Children never stop asking questions; unfortunately they are usually of the 'Can I go to the toilet?' and 'What do I do next?' kind! We need to model, look for and praise the questions that further enquiry, discussion and knowledge. The flexible thinking questions above combined with the 'What would happen if?' 'How do you know it is true?' 'What's the evidence?' 'What is your reasoning?' etc., will promote this culture of intelligent enquiry.

■ **Wonderment and curiosity:** Many children lose their natural curiosity once they have been in school for a few years. They stop noticing and asking questions; they look but they do not see! They will look at a tree, for example, but not notice that the leaves are different, that they seem to get smaller towards the top of the tree, and ask why?

We need to provide the children with wonder and amazement whenever possible; from swifts migrating to Africa without once touching the ground, to amazing records from the *Guinness Book of Records*, to mysterious numbers and beautiful literature! We need to ask the '**curiosity questions**' **ourselves**, the 'I wonder why …?', and hopefully the children will begin to copy our behaviour and share our enthusiasm.

Many of you reading this will recognise things that you do already! This does not surprise me, it is good teaching and common sense. However, we need to do this more systematically and consciously.

> We need to be more explicit in addressing these barriers to learning so that intelligent behaviour is more **taught** than **caught** and our children become independent thinkers and learners, which was the objective of the Big Idea at the beginning of this chapter.

REMEMBER: INTELLIGENCE; CAUGHT?; TAUGHT?

chapter 5

Use questioning and dialogue to lead children to understanding

The Big Idea here is nothing new, it goes back to Socrates and Aristotle and is part of teachers' everyday strategies and approaches. The first rule of the previous Big Idea was 'Allow children to talk', and in this chapter I want to explore this a little deeper.

There have been various movements in this area of education over the past few decades, from the Bullock Report in 1975 to the work of the Match and Mismatch project (Harlen et al. 1977) and Marie Clay (1979), to name but a few. The Education Reform Act of 1988 appeared to bring all these initiatives and many others, which were directed at higher orders of teaching and learning, including 'Thinking', to a premature halt.

It pleases me to say that there is today a movement which is not too dissimilar. It is based on the work of Robin Alexander: **Towards Dialogic Teaching: Rethinking Classroom Talk** (2006). This approach is being pioneered in projects in Barking and Dagenham in Essex and in North Yorkshire and goes under the more friendly title of **Talk for Learning**.

nb Dialogues, discussions, arguments and debates are at the higher end of the spectrum but this does not devalue the role in the learning process of rote, testing, recall, instruction and exposition by the teacher.

Hold intelligent conversations with children

Intelligent conversations are characterised in my simple terms as **'conversations which help children to find answers'**. This is what teachers aim to do every day. However, we perhaps need to be clearer in our own minds how to achieve this effectively. This section explores the above proposition.

These conversations can be with the whole class, small groups or individual children; the important factor here is the **quality of the conversation**. The one thing we know in teaching is that quality takes **time**! Earlier it was suggested that time was essential for children to think and respond; this is the same principle, although dare I say **teachers need time to respond**. If we are to work at the 'higher end' of the spectrum then we do need to put aside quality time and plan for it.

As we mentioned before, this conversation is a **dialogue** and it can be teacher led, or child led, or class led. The emphasis here is on the two-way nature of the conversation where **both the teacher and the children become good listeners**! This draws on many of the approaches that have been suggested before:

- Listening to others, not just hearing them while waiting to put your point of view.
- Trying to understand the other person's point of view ('OPV' as de Bono calls it in his **Thinking Course** (1997)).
- Valuing and respecting alternative points of view.
- Coming to judgements based on reasoning.
- Last but not least: learning to take turns and not to dominate the conversation.

Both teacher and children have to learn to **articulate clearly and coherently**:

- 'Engage brain before speaking', as one of my colleagues used to say to her children.
- Stop and think; decide which is the key Big Idea you want to get across; give an advance organiser and then take the other person(s) down a learning pathway to make your point. (Not too much different from lesson planning?)
- Children need to be taught this methodology and teachers need to model it.

This is not just social talk, it is **intelligent conversation**:

- Social talk is vital to the classroom; it is vital to oil the wheels and create an ethos of enjoyment.
- Intelligent conversation is distinguished by focusing more on the higher end of the spectrum (dialogues, discussions, arguments and debates).
- It encourages children to raise their own questions.
- It encourages children to challenge perceptions and avoid sweeping statements.
- It encourages children not to just 'hold opinions' but to 'justify their opinions'.

This conversation relies on **teachers asking intelligent questions**:

- Rubbish in: rubbish out! This was a favourite adage of computer programmers of the 1970s. (A very apt description of many of the radio and TV interviews that we are subjected to today.)

- We need to ask children questions which will further their thinking and learning. (Vygotsky: zones of proximal development, and scaffolding; 1978)
- Questions which further enquiry; exploration and explanation are particularly important in Primary science.

This conversation relies on **teachers giving intelligent answers**:

- Answers that further the learning: pose questions which direct children to rethink and not just give information, knowledge and fact.
- Answers which are **critically honest**. Praise and encouragement are vital to learning; however, if everything is praised then the currency is devalued. The correctness of the answer is not the issue, it is the **arriving at the answer that needs to be critically analysed** and the 'flaws' pointed out to the child(ren) in a positive and constructive way.

Hold scientific conversations with children

Scientific conversations are little different from the intelligent conversations described above; they are essentially dialogues that involve questions and suggestions which **help children to look for answers**, in science in this case.

The essential component and driver is as in the section above; it is **enquiry based**. The next section, on the Three Es, will explore this further, and above and beyond that, the specific approaches to Sc1. However, there are some general characteristics that permeate scientific conversations with children:

- **Vocabulary of reasoning:** This is the same logic as covered in the previous chapter on teaching children to become independent thinkers and learners. Children need to support their search for answers with argument based on logical reasoning. As you will see from a closer look at the KS2 SATs, many of the answers only require this thinking to be applied and no science, other than that given in the question. So this line of thought is of inherent value in itself.

- **Vocabulary of science:** This is another layer which needs to be added to children's thinking. Common sense tells us that there are many words which carry scientific meaning which are not used in everyday language, and if they are, they have a different meaning in science: e.g. dissolve, solution, inertia, momentum, heat, energy, genes, etc. Children need to acquire this vocabulary **in context**. This means it is best delivered embedded in the science they are learning, not separate from it. Simply choose two or three new scientific words each science lesson and explore, explain and commit to memory.

- **Evidence based:** This means that children need to be challenged to support what they say with **facts** and **evidence** and not unfounded supposition. Imposing this discipline will encourage them to stop and think and to take more care with their assertions. This will increase the probability of their being more accurate and less open to challenge and will equip them with a skill for life.

- **Open minded:** This is another intellectual skill with which children can be equipped for life, but which is sadly lacking in many adults regardless of their intellect or social status. It is the ability to look for and consider seriously other **alternatives** to the position you hold.

- **Arriving at judgements**: This is about coming to decisions based on the above reasoning and evidence and after considering alternatives. It is evidence based and open minded; it is arrived at on the **balance of probabilities** that it is 'more likely than not' to be correct.

Dialogues and intelligent conversations should not be uncommon in the 'enquiring classroom' of children of all ages.

REMEMBER: DIALOGUE; ARTICULATION; LISTENING

Teach children to enquire, explore and explain

Questions are the key to **learning science with understanding**.

Questions from the teacher and questions from the children, but not any questions in any order or the first thing that comes into the child's head; this may or may not be fruitful. What is needed is a way of working that removes that element of chance as much as possible.

Above is a **framework**, the Three Es, by which both teacher and children can **structure their thinking in science** and increase the 'fruitfulness'.

There are two obvious things to point out:

> 1   Teachers usually know the answers they are looking for.
> 2   Teachers usually prefer to draw the answers out from the children, rather than tell them the answers.

If we are to achieve 'learning with understanding' we need to use questions that **lead children where we want them to go**.

Below are some leading questions and their purposes:

Questions to promote enquiry, exploration and explanation

It's all about **children making connections** and drawing **inferences**, with the teacher's help. The three purposes above are about the **investigation process of science**. You may focus on one aspect of the process at a time or use them in a sequence through the investigation: beginning, middle, end.

1 Enquiry

At the beginning we need to direct children down the path of investigation that we want them to follow but without telling them directly. To achieve this we use questions like:

> I wonder if ... this will work better than this?
>
> I wonder if ... we can find out ... ?
>
> What do you think will happen if we ... ?
>
> Can you think of any way we could see ... ?

Sometimes it is useful to use a question which sets the scene and acts as an advance organiser:

> Which one will be **the best** floater (or runner or go furthest)?
>
> Can you find out which will be the best absorber (or conductor or magnifier)?
>
> Where does the grass grow the best? Why do you think it likes growing there? How can we see if we are right?
>
> Are all the leaves the same size and colour? How are they different?

2 Exploration

During the investigation we need to give children clues that will enable them to progress towards the outcome we are looking for and make sense of what they are doing as they go along:

> Did you notice ...?
>
> What else do you think we should try? This or that?
>
> Do you think that this will happen the same if we ...?

▶

> Do you think we could ...?
>
> How are we going to make this a fair test?
>
> How are we going to make sure that we can see clearly what the effect is?

3 Explanation

This is about helping children to make sense for themselves and bring out their natural talent for trying to explain the curious. It provides them with a reasoning strategy which will move them from **wild guesses** to **intelligent guesses**. Intelligent guesses show **reasoning** based on **inferences**: a chain of thought and the 'making connections' referred to above. Often it does not matter if it is a right answer; we can still praise the logic and then correct the misunderstanding:

> Can you think why this happened when ...?
>
> Could there have been any other reason ...?
>
> Why do you think this worked better than that ...?
>
> Can you think of another explanation?
>
> Can you think how we could have made it work better?
>
> Good thinking, I can see how you arrived at that; you first ..., then ..., and ... Smart move!

Don't forget the critical honesty:

> Nice try, but I think I can see where you are going wrong ...
>
> I'm not sure you are on the right track here; as Fagin would say, 'I think we'd better think it out again!'

nb Notice the use of the word **think**. This is far less threatening than asking a child **why**. Most of us are not sure of why, but if asked to share our thoughts we are only too willing. (An obvious statement perhaps, but one which escaped me as a beginning teacher until I read a book on Primary science edited by Wynne Harlen (Harlen 1985).)

These, then, are the **Three Es** which are generic categories to help us and our children approach Primary science in an enquiry and discovery mode. Sometimes, however, we need to think of questions which will promote particular science skills, and to build up a repertoire of them, which is our next rule.

REMEMBER: ENQUIRE; EXPLORE; EXPLAIN

Use questions to promote Sc1 skills

The key investigation skills that children need to practise are:

Observation

Looking for similarities and differences

Measuring and quantifying

Grouping and sorting

Looking for patterns

Fair testing

Controlling variables

Recording findings

Interpreting findings

These key skills are difficult to learn as we go along and as a result it can be better to teach them separately at the beginning (particularly in KS1), and as and when you see that the children need to practise a particular element.

You have probably come to the conclusion that there is a progression in the list of skills above regarding the intellectual challenge and that this is age related. The key here is, as has been said before in this book, that the teacher **gives** the children the other aspects of the investigation which are not the point of the exercise.

Observation

Direct children to **notice** and look closer:

How big/how small/colour/variance/detail? etc.

Comparison

Direct children to **compare**:

In what ways are they **different**?

Tell me what you think are the three main differences.

In what ways are they **similar**?

Tell me what you think are the three main similarities.

nb Notice that we chose on this occasion to ask for **a number of differences/similarities**. Sometimes it is better to be specific and ask for a manageable number, especially when the three that you have in mind will make your teaching point.

Does it remind you of anything else that you have seen before?

The Big Idea is to get children to try to make sense of what they are seeing, with your help.

Quantification

Putting **numbers** to things is very important in science. Children are naturally descriptive but do not readily think of **measuring** and **counting**. We need to promote quantification:

> How many?
>
> How big? Can we measure it? What do you suggest we use?
>
> How heavy? Can we measure it? What do you suggest we use?
>
> How fast? Can we measure it? What do you suggest we use?
>
> etc.

In the Early Years we can measure things approximately using **non-standard** measures such as matches, string, marbles, sand timers. Later we introduce the **standard units** of measurement, e.g. litres, kilograms, metres, minutes.

nb However, even though we have moved on to standard units, measurement at this stage of science is still **approximate**. Go to the nearest whole unit, don't go into decimal places! The Big Idea is that the results should clearly show the differences or patterns you want the children to discover, which is difficult if you are measuring to decimal places. If the results do not show obvious differentiation then **change the investigation so that they do**!

Grouping and pattern seeking

Part of the 'making sense' is to **sort**, **group** and **make connections**. This allows children to make the mental pictures which show them a meaningful **trend** or **pattern**:

> Can you sort them into groups for me?
>
> You can choose how you do it.
>
> Can you put them in order for me?

In the Early Years, children come up with ingenious criteria of their own. These are often personal and emotional and it is perfectly acceptable if they justify them in this way. Later we direct them to more scientific criteria which are physical properties and objective:

> Which have the most … and which have the least?
>
> Can you put them in order for me?
>
> Is there any relationship between the **size** of the leaf and the **place** where it grows on the tree?
>
> Does it make a difference to the **time** the salt takes to dissolve if you alter the **temperature**?

nb Notice that the two **variables** have been identified for the children. The point of the exercise is to look for patterns, not to control variables.

Fair testing

Children naturally have a good concept of 'fairness'; of advantage and, particularly, disadvantage. We need to direct them away from their emotional reasoning to **reasons** based on physical factors. (When testing toy cars to see which was the fastest one of my children was convinced it would be the 'red one', because his Dad had a red car and it was real fast!)

It is sometimes helpful to make the investigation **grossly unfair** to certain children, who will then jump on you like a ton of bricks, and then the point is made.

The Big Idea is simple: to be fair, all the test conditions must be the same for each contestant, be they animal, vegetable or mineral.

Controlling variables

This is probably the most challenging aspect for teachers to teach and children to learn.

The Big Idea is that we are trying to see and measure **action** and **reaction**. In order to do this we have to make sure that there is a clear **one-to-one correlation** between what we do and what happens as a result of this action.

Science is all about action and reaction!

In science, things that alter are called **variables**, we need to control these so that **only two factors** are changing: one that **we experimentally alter**, which is known as the **independent variable**, and the other factor which alters **as a result of that action** we have just taken. This is known as the **dependent variable**.

nb When **plotting graphs** of the results, the independent variable goes on the horizontal x-axis and the dependent variable on the vertical y-axis.

> What are **we** going to change? (Variable 1: independent) **Action**
>
> What effect do you think we will see?
>
> Will it affect more than one thing?
>
> How can we make sure that only one thing changes as a result of our actions? (Variable 2: dependent) **Reaction**

Recording

The Big Idea here is to capture the findings in some form of 'hard copy' so that we can analyse and interpret them to see what they mean.

Descriptive recording

There is a progression through the Early Years from teacher script and pictures to narrative and more formal drawings and diagrams.

Measurement and counting

Number

Order

Grouping

Tally charts, pie charts, e.g. (Q3a KS2 Test A 2007 Science: Teeth)

Matrix charts, e.g. (Q6d KS2 Test A 2008 Science: Bananas)

e.g. (Q7b KS2 Test A 2008 Science: Floating)

e.g. (Q6a KS2 Test B 2008 Science: School pond)

e.g. (Q8b KS2 Test B 2008 Science: Parachutes)

Bar charts   e.g. (Q3b KS2 Test A 2007 Science: Teeth)

Line graphs   e.g. (Q9c KS2 Test A 2008 Science: Candle)

e.g. (Q4d KS2 Test B 2008 Science: Hot drinks)

Binary keys   e.g. (Q2a KS2 Test B 2008 Science: Penguins)

At first it is necessary for the teacher to produce the chart/graphs and put in the headings of the variables and the scale and units on the axes if appropriate. With experience, children will begin to make their own decisions on how best to record their findings.

> Which would show our results better, a bar chart or a line graph?
>
> How can we find out? (A rough sketch will usually do the trick.)

Interpreting

This is the key opportunity for the teacher to develop the children's **reasoning** skills; to help them to see the **connections** between action and reaction and then to try to **explain what they might mean**.

Common-sense explanations are perfectly acceptable, especially at the beginning, and it is up to the teacher to tease out the child's logic. Later on we are looking for the **appliance of science** in our explanations, i.e. some scientific knowledge should be included in the explanation if it is appropriate.

It is also an opportunity to make connections not only with further science, but also other subjects such as Technology, History, Geography.

✱ eg Friction is a helpful force when we want to stop, but very unhelpful if we want to keep moving! That's why we lubricate our engines and wheels with oil so that they don't get worn away with the friction.

nb Remember the emphasis on the 'what do you think?' Do not be afraid to give children clues and choices to help them arrive at the explanation. See the NB at the end of the Three E's section above (page 89).

chapter 6

You need to know what children have learnt and understood before you progress

This Big Idea is another statement of the obvious; however, it is easier said than done. Assessment can be a very time-consuming and bureaucratic procedure if we are not careful; on the other hand it is essential to the teaching and learning process.

There is a great deal written about assessment: formative, summative, diagnostic, **of** learning, **for** learning etc. This is important and valid and is trying to address the Big Idea we have just posited. We need to answer the following questions:

Where are my children now?

How do I know?

How do they compare with other children who are similar?

Where do we need to go from here? (Progression)

nb Progression is not 'more of the same but a little harder'!

It is about deepening as well as widening the children's knowledge. It is about the child's progress to becoming an independent thinker and learner of not only the content of science, but also the process of science.

If we want to know what progress the children have made to date then we have to **assess** their progress to date.

Make reliable assessments of the children's achievements

This is a measurement of what children **know, can do** and **understand**, in terms of the NC Programme of Study. However, to make it reliable it needs to be **objective** and set against agreed **criteria**.

The objective criteria are the **levels of attainment** laid down in the NC attainment targets. The only problem is the need to ensure that all teachers **interpret them in the same way**. This is achieved to a greater or lesser degree by giving teachers exemplars and by internal and external moderation between teachers and between different schools.

Guidance and exemplars are provided by the QCA to try to provide consistency across the country and at KS2 the statistical data should corroborate the teachers' assessments. This process is being refined by the QCA to provide a simpler and more reliable approach to assessing the attainment levels in Science and which is due to be published. It will still be to national standards but will replace the level guidance in the National Curriculum and be modelled on the same lines as the Assessment of Pupil Progress (APP) in Literacy and Numeracy.

The model that is used is that each level is analysed into a progression of six foci. As the child completes the progression the level is achieved. The model will take the form of a matrix with the six foci heading the six columns, and statements of progress will be made underneath each one in sub-levels of increasing demand. Teachers will be able to highlight the level in the focus that the child has reached. The **overview** will be the achievement of that particular level in that particular attainment target. This will be accompanied by guidelines, exemplars, a standards file and a handbook.

This new model will enable teachers to arrive at more robust decisions and, if the correlation between the teacher assessment and the SAT is statistically reliable, then who knows what might happen in the future to the KS2 SATs?

REMEMBER: KNOW; DO; UNDERSTAND

Don't run before you can walk

The advice from the QCA is to try the instrument out on a small sample of children first to familiarise yourself with the criteria, before trying to assess the whole of your class. This will then provide you with the confidence to look at the children individually.

An alternative approach could be to carry out the assessment process **deductively**, i.e. go from the general to the particular. In other words, **start at the class level of assessment** and assess the strengths and weaknesses of the class as a collective not as individuals.

■ Assess the **class as a whole** and get an overview of the class's strengths and weaknesses, i.e. use **one** sheet for the general scientific abilities of the whole class at that **level** and that **focus**. In this way you can get an impressionistic overview which can inform how you plan your science to meet the class's needs.

> **✱ eg** This might show you that the majority of the class are reluctant to begin to explore, even when given help. This independence is a key aspect of L3 Focus.

This information would then direct you as the teacher to provide plenty of 'hands-on' activities to build the class's skills and confidence in this area.

■ In the same way you could use this approach as a performance indicator for particular groups, i.e. **one** sheet per group, so that you can begin to differentiate activities for the class, intelligently.

- In the final case you are going to have to arrive at a full assessment of each child's level at the end of the year, which does mean generating an **individual** assessment sheet. However, based on the previous class and group assessments, this could be already half filled in.

This really brings us to the next Big Idea ...

chapter 7

Make assessment work for you and your children

Accountability is vital in teaching; it is the process that determines the effectiveness of the teaching and learning from class level, to school level, to local authority level and, most importantly, gives parents and carers the measure of their children's success and their teacher's success, not to mention the accountability to the DCSF and the taxpayers!

However, to use assessment just for bureaucratic purposes alone would be a tragedy! We need to use it to inform 'School Improvement', which was the movement introduced in the 1990s by the DfES to drive the government's education agenda and raise standards. School Improvement is still the challenge that has to be met by every headteacher and their board of governors, which means that data should principally inform not league tables but the School Improvement Plan for Achievement and Attainment.

Schools are 'data rich' and the plethora of information that is provided on 'Raiseonline' by the Standards Unit can provide very useful intelligence for the senior management regarding the school's attainment and inform the School Improvement Plan. This analysis shows the school's trends of attainment over the previous 3–5 years and yearly analysis of performance, which, when combined with age-related expectations, provides the school with a tool to monitor and track pupil progress and the value added. This evidence then provides sound reasoning for intervention and enrichment to address the problem areas.

Don't just record it; use it

It has been said before that if class teachers are making records that are only to please their line managers and, in turn, their line manager, then seriously question the practice. Records of assessment should provide **intelligence to inform your practice**.

Below is a **suggested** line of reasoning that may help you to examine the assessment data that you gather from your class and provide you with the 'intelligence' to teach with more precision:

- Assessment, whether impressionistic or captured in hard copy (test results, marked work, observation notes e.g. QCA, APP level indicators), is usually only a **sample** at one point in time.
- To check the **validity** of these findings we need to repeat the assessment process and gather further data on other occasions.
- Accumulate the data and record it alongside previous data.
- Sort and group the data so that the reorganisation makes it easier to compare, and to see differences and similarities. (Ironically this is a skill which we teach in KS1 Primary science!) Not just alphabetically or rank order, look for **clusters**.
- Analyse the data by **class**, by **group** (ability), by **child**.
- Further analyse the above data by other categories, e.g. by attainment target and by level.
- Interpret; look for **trends** and **patterns** in the data; also look for **issues** and **dilemmas** that are thrown up. (Highlight and annotate your records to provide an aide-memoire.)
- Lastly, compare findings with previous findings to arrive at **trend data** that shows consistency over the term and the

▶

year in a particular class or Year Group and from year to year between classes and Year Groups. (It is more likely that the Science coordinator would be interested in the latter.)

The above methodology will help to show you **progress or the lack of it**. We must not forget that although the focus of this book is children's understanding of science and a possible change in pedagogy, the book is also about **improvement in achievement**, by which we mean levels of attainment since this is the yardstick that we are given by the QCA to measure achievement.

Alongside the absolute measure of attainment required for purposes of **accountability** and fair local and national comparison, the process of measuring that attainment should generate the **intelligence that informs our teaching and the children's progress**. It could show 'flat-lining' for individuals or groups or the whole class and the need for further remediation and enrichment. On the other hand, it could show complacency and a need for challenge and extension.

More importantly it will direct you to the **area of need** in the Programme of Study so that you are teaching to the required target to inform the improvement of science achievement.

REMEMBER: RECORD; USE; INFORM

Use impressionistic teacher assessment; it is valuable

Throughout the day, teachers receive and process hundreds of impressions about their children's work and behaviour. The problem is that these tend to be rather ephemeral and disappear in the hurly-burly of classroom interaction. Rather than just absorbing as we move around the room we need to reflect later and be a little more analytical and note the patterns, similarities and differences in a more conscious way than perhaps we do at the moment. This can often tell us things about our children and our class that we did not suspect.

Most teachers are sensitive to these signals from the children and almost subconsciously build up an impression which gives them an instinctive feel for the class. The argument here is for teachers to test that impression more analytically.

It can also be useful to be more **proactive**. Since you are the teacher you already know the specific learning outcomes which you are looking for, so look for those specifically. In other words, 'Have they got it?' This is something that teachers do all the time, but often in a more episodic than methodical way. At this point we have moved beyond impressions and need to focus on the specifics rather than the general. In this way you have begun to refine the problem deductively from the general and started to identify particular outcomes; now the attainment problem will seem more manageable and less overwhelming!

nb Both casual observation and focused observation are valuable in their own rights.

nb Since the QCA has provided us with foci to observe against, which are consistent in Primary schools across the country, it might be sensible to use them!

Remember: assessment is important to authorities, but it is more important to teachers and their pupils.

Science knowledge

The 'Big Ideas' of Primary science

The philosophy of Big Ideas is based on the theory that if you give a child the big picture first as an advance organiser, then they will be able to make more sense of the individual lesson inputs. It is a way of simplifying science without dumbing it down and relying solely on 'tips for teaching'.

All knowledge can be regarded as a hierarchy of ideas which begins with the most general and progresses to the particular. Over the years this deductive process has moved from the study of philosophy, which was a combination of language, arts and science, to not only biology, physics and chemistry but also to sub-divisions within those sciences, e.g. particle physics and molecular biology.

In this section we shall try to return to the simple Big Ideas that underpin the three core sciences at this level. In the eyes of a 'pure' scientist the ideas may not be scientifically accurate; however, at this level we are trying to deliver a concept that readily makes sense and is more or less scientifically correct. The exceptions and anomalies are not dealt with at this level; they will come as the student progresses with their science education. Our job is to 'turn them on' to science; to complicate matters would appear to me to be counter-productive, e.g. to complicate Light by introducing photon theory, or to introduce relativity to challenge Newtonian physics would be unnecessary at this age.

This section reduces Biology, Chemistry and Physics to a series of Big Science Ideas contained in the NC PoS. Alongside the idea there is a common-sense simplified explanation of the science. The reader may wish to refer to other Primary science books which specifically address the science content and its understanding.

chapter 8

Sc2 Biology

It is well known that Biology is an 'ology', but unlike Physics and Chemistry it is about living things and at the Primary level it is important to ground it in **nature** and not in the laboratory.

The Big Science Ideas of Biology

Big Idea	Notes and further explanation

Life processes:

- Be it plant, animal or microbe, it is all about **living, growing and reproducing our species.**

  Feeding and not being the food of others.

  Designing an efficient method of reproduction which ensures fertilisation and offspring in greater or lesser numbers.

- This process is a **cycle**: birth, growth, life, reproducing and dying.

  This cycle is mirrored by breaking down of the dead matter and then the recycling of the essential biochemicals from which the living form can be synthesised, e.g. carbon, hydrogen and nitrogen.

- In order to successfully achieve the above cycle, each plant or animal species has developed its own unique way, which is adapted to its environment.

  It is important to see the same principle in action for all living things, and not only its similarities but also the wonder and variety of differences.

Humans and other animals:

- Food input and the need to break it down to release energy and vital biochemicals for rebuilding the body's structure.

  It is called 'ingestion and digestion'.

  All animals have various ways of physically crushing and grinding the food into smaller pieces and then dissolving it.

▶

Big Idea	Notes and further explanation
■ To maintain our health we need to eat a healthy diet and look after our teeth.	All animals have a variety of ways internally or externally of secreting chemicals to break down the material and obtain the essential biochemicals which are the building blocks of life.
■ Blood is the liquid medium that is used to carry oxygen and all the other vital chemicals to every cell in the body.	Molecules in the bloodstream can attach to other molecules and act as a carrier to take them to wherever they are needed in the body. Oxygen is attached to the red blood cells. Carbohydrate (CHO) molecules, such as starches and sugars, can be taken and stored in cells and later broken down by other chemicals (enzymes), to release the energy vital for life.
■ The blood is the vital first defence of the body against infection.	It carries white blood cells which mop up infection.
■ To provide the force to achieve this there is a 'heart pump' at the centre of the operation.	This system is a 'closed' system, i.e. it has an outward pathway through the arteries and a return route via the veins.
■ The more energetic we are the harder the pump has to work to supply the increase in demand.	The arteries immediately provide a supply back to the heart muscles to supply the massive energy demand of the heart muscles before supplying any other muscles.

Big Idea	Notes and further explanation

The next essential demand on the vessels is to take blood to the lungs where it releases waste carbon dioxide and picks up oxygen to take around the body to release the vital energy of life.

■ The heart pump does the same job for all animals but differs in its form and structure.	Not all animal hearts are the same design, e.g. number of chambers, but they all perform the same function.
■ Locomotion is important for all animals, some are very fast, and others are very slow.	Speed is not always the most important factor, being adapted to your environment to secure your particular food supply, is!
■ Locomotion is helpful in acquiring food and avoiding being eaten by other animals who are after your biochemicals.	Animals are ingenious in finding and catching their food, from hunting solo, to packs to whole armies (ants)!
	Speed is important in avoiding predators, but some animals have other wondrous ways like camouflage and only exposing themselves at safer times.
■ To move, animals have to overcome the force of gravity.	Gravity is the limiting factor of size. The larger and hence heavier you are, the slower you move and so become more vulnerable to the sleeker and faster species.
■ To do this some animals have developed skeletons and muscles, either internally or externally.	If you are big then you need very good defences, such as a thick skin or armour or a defensive weapon, e.g. tusks, horns or poison.

▶

Big Idea	Notes and further explanation
■ Other animals have adapted to living in the water where the force of gravity is felt less.	This could explain why life began in the water as the gravity problem was easier to solve here.
■ To remain healthy the body has to excrete the toxins we take in with our food and drink.	Most chemicals in excess are dangerous, even salt and sugar, so the body removes the excesses via the liver and kidneys and through the pores of the skin.
■ Particular toxins can seriously damage and eventually kill you, e.g. tobacco, alcohol, drugs.	If the body cannot break the poison down and excrete it quickly enough it will make you ill. Even pharmaceutical products are toxic and have side effects that are poisonous to a slight degree and can kill if over dosed.
■ The body is a machine and we need to keep it 'well oiled' and well used or it will 'rust' and fail.	If you do not expend as much energy as the food energy you take in, then you will tend to put on weight and become unfit. If you do not use your muscles and walk and run, then your muscles waste and your bones weaken, which can affect you later in life.

Green plants:

Big Idea	Notes and further explanation
■ Plants also need food, water and air, but obtain it in different ways from animals.	Plants usually have roots to anchor them to, and obtain water and nutrients from, the ground. They usually have a stem structure which is strong enough to overcome gravity and enable them to grow upwards and compete for light without which they will die.

Big Idea	Notes and further explanation
■ Unlike animals, and almost uniquely, green plants are the only living things that can capture the sun's energy.	Plants not only capture the energy in the sun's light, they also absorb carbon dioxide and combine it with water to make energy-containing food (carbohydrate molecules such as sugar and starch). This food is a source of energy and growth not only for themselves but also for the animals that eat the plants.
	By a fantastic coincidence the waste product that plants release is oxygen, which is vital for all animal life! So plants take our waste product (carbon dioxide) and give us theirs (oxygen) in return. Neat!
■ The energy from the Sun's light is the source of all life on the earth.	Green plants are therefore the original source of food for nearly all the living species in the world.
■ Plants cannot get up and go and meet their mate, so they have come up with ingenious ways of reproducing themselves and dispersing their offspring to give them the best chance of survival.	Plants have male and female parts just as animals do. The idea is for them to meet and fertilise the egg to produce new offspring. Since plants cannot move they have to use ingenious ways of transporting the male pollen to the female ovary (insects, wind, water).

▶

Big Idea	Notes and further explanation

By the same token plants need to disperse their offspring if they are not going to fall to the ground around them and all compete for the same food supply. To do this they pack them up as seeds with their own food supply so they can germinate on their own. They then equip them with ways of being carried away by the wind, water or animals or being eaten and later excreted (fruits and berries). Clever!

Variation and classification:

■ There are thousands of plant and animal species and if we are going to get to know them we first have to organise them into some kind of logical order.

The eighteenth and nineteenth centuries produced lots of collectors who were keen to discover new species and have them named after themselves.

■ To do this we look at them carefully and put similar ones in the same groups.

The eighteenth-century Swedish naturalist Linnaeus, invented a system of naming and grouping plants and animals which would later become universally recognised. Unfortunately for some of us it is based on their scientific names which are in Latin.

Big Idea	Notes and further explanation

In order to name them we use identification 'keys' which trace the plant or animal and give us their correct name. (These keys are known as 'flora' for plants and 'fauna' for animals.)

The principle of the key is that it asks you a question to which you may only answer yes or no. (Binary, which is how in principle the computer works.) If the answer is yes, you follow one route, if the answer is no, you follow another route, and so eventually arrive at the picture of the plant or animal that you are looking for.

Living things and their environment:

All living things are precious and we need to treat them with respect and protect their environment so they can live and grow in peace.

All living things live in balance with other living things. In order to survive they adapt to a particular way of life in a particular environment. If we destroy that environment we destroy the living things that live in it.

Plants and animals live in particular environmental habitats to which they have become adapted.

Plants and animals take different forms in order to live grow and survive in different habitats, from land, sea and air, to extreme heat and extreme cold.

The original source of food on our planet is the green plant. If animals are to survive they have to eat green plants, or eat other animals that have already eaten green plants.

The building blocks of life are the chemicals which combine together to form the molecules which construct the billions of cells that make up the plant or animal.

▶

Big Idea	Notes and further explanation

In order to carry out this process we need to make even more complex chemicals (**enzymes**), which will facilitate the biochemical reactions to live, breathe, grow and reproduce. We eat to obtain these essential chemicals.

Since these reactions occur in a watery medium, plants and animals will die if they cannot get this water. Most plants and animals are made up mainly of water and can only last a matter of days without it.

■ All life is competing with all other life for food. The species that are best adapted to their environment will survive and carry on reproducing and populating their habitat.

This is the principle of **evolution** and **natural selection**, i.e. adaptation and survival of the fittest (**Darwinism**).

■ In this fight for life you are either a **predator** or the **prey**; however, most animals are in both roles simultaneously!

Animals eat other animals and plants, but at the same time they have to prevent themselves becoming food for other animals who want to eat them.

Animals and plants become very clever at disguising and protecting themselves from predators.

■ Some species are 'top dogs' and at the top of this **food chain**; others are below and are fodder for those above them.

We can organise these feeding relationships into food chains which help us to understand the fragile balance of life in a particular environment.

Big Idea	Notes and further explanation
■ Green plants are obviously at the beginning of this chain and lie at the bottom.	Since the green plant is really the only life form that can capture the sun's energy, then life has to begin here. (Plankton in the sea and some bacteria also serve the same purpose.)
■ Micro-organisms are so small that they cannot be seen with the naked eye.	We were not fully aware of them until the microscope was discovered and we could see them. Electron microscopes have now been invented which allow us to see inside the cells themselves that make up all living things, both micro and macro.
■ Micro-organisms are living organisms and essential in the decomposition process, breaking down plant and animal waste to release the chemical building blocks we need to acquire to live and grow.	There are millions of 'good bacteria' which work inside our bodies, helping us to break our food down, and outside our bodies breaking down dead plants and animals and returning the vital chemicals to the earth to complete the **recycling** process.
■ Unfortunately as well as being vitally helpful, they can also be harmful and deadly!	On the other hand there are harmful bacteria and viruses which invade our cells and threaten life. We are constantly developing chemicals which will kill them but will not harm us (e.g. penicillin and other antibiotics).

chapter 9

Sc3 Chemistry

Chemistry is the study of particles, known as atomic elements, and their molecules, which make up all materials both living and non-living. Individual elements contain their own unique atom, whereas the molecules are formed by combining different kinds of atomic elements. There are over 100 different elements which are known to occur on the Earth and these are organised by their chemical properties in the **periodic table** (created by Mendeleev in 1869).

The Big Science Ideas of Chemistry

Big Idea	Notes and further explanation

Grouping and classifying materials:

- All materials have different properties owing to the different chemicals that make them up.

    Some conduct electricity, others do not. Some are magnetic, others are not. Some are hard; some are flexible; some are solid others are not, etc.

- Materials with similar properties are usually classified together to help us make sense of Chemistry, in the same way as we did in Biology. Other materials are classified by their appearance and texture.

    The properties decide their everyday usefulness and this provides us with a rough classification, e.g. wood, glass, plastic, metal, stone, brick.

    It can be useful to group those that are good conductors of heat or electricity, or not so good, i.e. insulators.

    Ultimately materials are classified according to their chemical make-up. This basically falls into two categories:

    — Those based on the **carbon** atom (**organic chemicals**, e.g. hydrocarbons such as oil and food stuffs).

    — Those **not** based on the **carbon** atom (**inorganic chemicals**, e.g. water, iron, copper, sulphur etc.).

▶

Big Idea	Notes and further explanation

■ Materials occur in one of 'three states of matter': **solid, liquid, gas.**

Its state depends on whether the atoms and molecules are:

— packed together closely and unable to move, as in the solid state, or

— more loosely connected and able to slide around each other and change shape, as in the liquid state, or

— can break free from each other and fly around, as in the gaseous state.

■ If you input energy then material can change state to the higher form, i.e. from solid to liquid to gas. If you take energy out then they will revert to the lower form.

The states of matter are **decided by how much energy they have**.

Inputting energy allows them to begin to break free from the **chemical bonds** that hold them as solids and move into a liquid phase and eventually break completely free and move into a gaseous phase.

Changing materials:

■ The simplest changes occur when materials are mixed together.

■ Mixtures are materials that keep their separate identity.

Some materials when mixed together do not change but stay separate, e.g. sand and salt, or water and oil, or sand and water.

■ Other solid materials when mixed with some liquids will lose their identity in that they disappear into the liquid and are hidden, but can be retrieved if the liquid is driven off.

Other materials **dissolve** into other liquids, which are known as **solvents**. The most common solvent is water; however, there are powerful organic solvents such as alcohol or the trichloroethylene used by dry cleaners.

Big Idea	Notes and further explanation
■ Most chemicals will react with other chemicals if energy is released from inside the molecules or added in the form of heat. In the same way a change will occur if energy is taken away.	Materials remain stable when their energy system is in balance. If this is upset by the addition of, or subtraction of, energy, then change occurs.
■ Water is the most common substance on the earth's surface and when heated or cooled it changes state.	The simplest changes result in a change of state. The change of state from water to steam is called **evaporation**; the reverse reaction is called **condensation**. **Temperature** is the scale with which we measure the amount of heat or cold.
■ All the water in the world eventually changes state and is recycled.	Water evaporates from rivers, lakes and seas to form clouds. These then cool and condense and return the water to the earth's surface which eventually flows into the seas. This water is not pure as it picks up chemicals from the land as it flows down to the sea. These chemicals make up 3% of the sea's mass, which means there are millions of tons of gold particles in the sea! (Not to mention all the rubbish man dumps into the water!) Recently, due to global warming effects, even the ice from the poles is joining in this water cycle.

▶

Big Idea	Notes and further explanation

■ Chemical changes are known as **reactions**, which can proceed one way and, under certain circumstances, can change back, in which case they are known as **reversible reactions**.

Reactions occur when combining chemicals upset each other's **equilibrium**. When this occurs the law of nature tries to overcome this destability by taking the energy from one of the reactants, resulting in a chemical change to a more stable state and producing a different material.

■ Some **non-reversible reactions** produce new materials which can prove to be very useful.

Sometimes this results in the creation of a completely different material, e.g. sand, cement and water. Most new materials, from stainless steel to PVC, have been discovered in this way.

■ When materials are combined with oxygen and heat, they burn and produce new materials which are not normally reversible.

This process is known as **combustion** and results in the ignition of the products produced.

The energy released from inside chemicals in this reaction can be considerable and, if constricted and released quickly, can cause explosions.

Separating mixtures of materials:

There are four basic chemical separation techniques:

■ Sieving

When there are **mixtures of solids** which are of different particle size, we can strain them through a sieve and the smaller particles will pass through the holes, leaving the larger ones behind.

Big Idea	Notes and further explanation
■ Filtering	When there is a **mixture of an insoluble solid in a liquid**, we can use the same sieving principle to **recover the solid**. In this case the holes are the minute pores in special filter paper which let only the liquid through. Cloth and tights will also act as a cruder filtering system.
■ Evaporating	When there is a **mixture of dissolved solid in a liquid** we can drive off the liquid using heat to change it into the gaseous state, **leaving the solid behind**. The most common liquid is water, and since eventually everything ends up in the sea, we could recover everything from salt (pans) and other soluble materials, to filtering out the millions of tons of gold particles that are hidden there.
■ Condensing	When there is a mixture of **dissolved solid in a liquid** and we wish to **recover the liquid**, not the solid, then we can use the same principle. This time we catch the hot gas in a **cool** chamber and it will **liquefy** and fall to the bottom of the glass. This is an excellent way of producing pure water and is the principle of the desalination plant.

▶

Big Idea	Notes and further explanation
	The same principle is used to **separate liquid mixtures**, e.g. crude oil contains many oil fractions, one of which is petrol, and each has a different boiling point. As a result, the crude oil can be carefully heated up and, as the fractions rise separately, they can then be tapped off and **cooled and condensed** into separate containers. This process is known as **fractional distillation**.

chapter 10

Sc4 Physics

Physics is the study of physical forces and laws of matter. They range from the sub-atomic nature of matter and the forces that hold it together, to the actions and reactions of that matter in the universe. At Primary level these are classified into electricity; forces and motion; light and sound; and the Earth and beyond. We are seeking to understand what makes things happen on the Earth and beyond, and what are the laws and principles that govern these phenomena.

The Big Science Ideas of Physics

Big Idea	Notes and further explanation

Electricity:

- Electricity is a **form of energy**, as is chemical energy or energy of movement or light energy or sound energy or heat energy or nuclear energy.

- Energy can change from one form to another form.

Electrical energy can be seen in nature as lightning or be made at power stations.

Power stations take the energy trapped in coal, oil, gas or in nuclear reactions and convert it into the electrical form of energy.

- Electricity is dangerous.

Electrical energy from the power station to our homes is dangerous and **kills**!

Electricity can be made and stored **safely** in **batteries**.

- Because electricity has energy it can do very useful work for us such as powering our electrical appliances and turning electrical motors to drive machines, not to mention lighting up our planet and keeping us warm.

Electricity is a form of **energy**, which is a word that describes a force that makes us and other things **move** and **work**.

Electricity is used to power most of the world's devices, from those in the home to the world of work. This is because it can be conducted to where it is needed at lightning speed, literally. (Beats steam any day of the week, which was the source of power before electricity was 'discovered' in the nineteenth century.)

▶

Big Idea	Notes and further explanation
■ Electricity is made up of billions of electrons.	This electrical current depends on the force which is pushing it along. This force is measured in **volts** and this is proportionate to the difference between the high point and the low point, which is known as the **potential difference**. The amount of electrons carried in the current is measured in **amps**. Anything that tries to stop the current flowing is known as a **resistance** and is measured in **ohms**. All these three factors affecting electrical flow are interdependent and this relationship is expressed in a law called 'Ohms law'.
■ In order for these electrons to flow they need to be able to return to where they started from, i.e. follow a return route or go to 'earth', literally.	This is the phenomenon of **completing a circuit** and is what happens to us when we accidentally touch something **live** and get electrocuted!
■ If we want to stop electricity flowing we simply have to break the circuit.	A **switch** is the device which is used to interrupt the flow of electrons and is either **on** or **off**. This binary property of electricity is the principle of the computer, sort of like Morse code (dots and dashes which can be translated into words and numbers).

Big Idea	Notes and further explanation

Only when the electricity is in the **on position** will it flow and give up its energy to power things. When it is still, it is harmless.

Static electricity is electrons found in nature, not in 'wires', which are still and harmless. However, when they are given an 'escape route' they shoot down it at great speed, releasing their energy as they go. This is the principle of lightning and sparks and shocks when removing synthetic clothes or touching the car door frame.

■ These electrons can travel very quickly through some materials but not through others.

Electrons can travel with very little resistance through some materials, e.g. metals like copper, steel and aluminium. These are known as **conductors**. Other materials such as plastic, rubber and wood, provide such a great resistance that the electrons stop flowing. These are known as **non-conductors**.

▶

Big Idea	Notes and further explanation
	Primary science circuits are powered by a safe source of electrons which can be found in batteries. The rest of the circuit needs to be completed by a conducting circular route back to the other side of the battery. This route normally takes the form of **wires**, which are covered in a non-conductor such as plastic to stop the electrons escaping. The wires are usually copper because this metal is a very good conductor and not as expensive as the best conductor, which is gold.
■ If we want electricity to do work for us then we simply have to add the device into the circuit and the electrons will pass through it and give up their energy to power that device.	When the electrons pass through devices such as light bulbs, buzzers and motors, they make them work!
	The amount of electricity flowing can be increased or decreased by the power available (adding or subtracting batteries or using batteries with a higher voltage).
	The more devices we have in the circuit then the less electricity each one receives and so the circuit does not work very effectively.
	Sometimes, how we connect the devices into the circuit allows the electricity to flow better and work more effectively (**series** versus **parallel**).

Big Idea	Notes and further explanation
◼ Electricity and magnetism are very closely related.	When electricity flows it creates a magnetic field around it. Conversely, if we create a magnetic field around an electrical wire then the electrons begin to flow along the wire creating an electrical current.
	This reciprocal relationship of electricity creating magnetism and magnetism creating electricity, is the principle behind motors, dynamos, electromagnets, remote door-locking devices, speakers and devices which convert electricity to sound and vice versa, e.g. telephones, TV and radios.
◼ Since it is difficult to visualise and remember electrical circuits, it is helpful to draw them.	So that we can all read each other's circuit diagrams we all need to use the same symbols and format.

Forces and motion:

◼ Certain materials known as **magnetic**, exert an invisible force which causes them to be attracted to or repelled from each other.	Ferrous metals such as iron and steel, plus cobalt and nickel are the only things to exhibit magnetism.
	These magnetic forces produce a north and a south pole when they become lined up.
	Like poles repel and opposite poles attract.
	Because these elements are in the core of the Earth our planet has a magnetic North Pole and South Pole.

▶

Big Idea	Notes and further explanation

- The Earth is affected by a massive force known as **gravity**. This force acts all over the Earth and is a downward **force** which pulls everything down to the surface.

Gravity is a massive downward force acting to pull things down to the centre of the Earth. That's why if you fall to the ground at **speed** and that causes you to **crash** into the surface with sufficient force, you may be killed. It is also the reason why, if you want to leave the surface, you need a bigger force acting in the opposite direction, e.g. jumping or flying.

- **Gravity** is an invisible and little understood force that is a property of matter which causes other matter to be attracted to it. The **denser** the matter then the greater the force of attraction.

The amount of gravity an object exerts depends on how closely together the matter in it is packed together. The more densely it is packed together then the **heavier it will be for its size**. This is called **density** and depends on the mass and how much of it there is in a set volume, i.e.

Density = mass/volume

- Measurement is critical in science and to be fair we have to use the same scientific units of measurement all over the world.
- Gravity is a force and is measured in **Newtons**.

Because of the Earth's density, the force of gravity is approximately **10 Newtons** all over the surface. (A **Newton** is the **unit** used to measure **force**.) Denser planets have higher forces of gravity, e.g. Jupiter. The less dense then the less the force, e.g. our Moon is one-sixth as dense as the Earth and exerts one-sixth the force of gravity, hence the 'bouncing astronauts'!

Big Idea	Notes and further explanation

- This gravitational force of attraction between massive heavenly bodies is what 'ties' them together.

The Sun's gravity is the force that keeps the solar system in orbit and stops it from flying off into outer space!

- The force of gravity on an object on the Earth's surface is commonly felt as its **weight**.

Weight is what results when gravity pulls down an object to the centre of the Earth and because this is a measure of **force** its unit of measurement is **Newtons**.

- **Weight** and **mass** are not technically the same and are measured in different scientific units.

The 'stuff' that makes up the object is known as its **mass** and is measured in **kilogrammes**. This is confusing for two reasons:

1 because we 'weigh' goods in kilogrammes (or pounds), when scientifically we should weigh them in Newtons

- Away from gravity there is no weight, e.g. astronauts in space.

2 because on the Earth's surface weight and mass appear the same; however, in space away from gravity we would weigh nothing but still register our 'stuff' (mass) in kilogrammes, unless we had evaporated!

- When a **force acts** it results in things **moving**, or being made to **stop moving**, or being made to **stand still**.

The direction of the force determines which direction the object will move in. If the **force increases** the speed of travel will increase; this is known as **acceleration**.

▶

Big Idea	Notes and further explanation

- Besides needing to **measure** the size of the force we also need to know its **direction** of travel.

The opposite effect will occur if the **force decreases** or the **force acting against** the direction of movement increases; this is known as **deceleration**.

- In Physics we refer to **speed** meaning the amount of distance travelled in time (d/t), e.g. miles per hour or metres per second.

- **Velocity** is technically different in that it is measured in distance, time and **direction** and so must include a vector arrow showing direction. (Confusing because both speed and velocity generate the same numerical value, i.e. d/t.)

- **Acceleration** is the change of speed or velocity over time and so is measured as speed/time, the unit of which is speed per second; which is metres per second per second or metres per second squared!

If the **force** continues to **stay the same** then the object will keep travelling at the **same speed** all the time, not getting any faster or any slower. This is known as **uniform speed** (or uniform velocity if you give direction). It is very difficult to achieve unless you have 'cruise control' on your car. We are always accelerating or decelerating whether we travel on land, sea or in the air!

Newton discovered three laws that explained motion:

1 Nothing moves or stops moving unless a force makes it.

2 The bigger and heavier an object is then the more force is required to move it or stop it.

3 For every action there is an equal and opposite reaction.

These are my words! The last law was genius and appears counter-intuitive, but just think about rockets: if we want them to go up we push a force out of the back, downwards!

Big Idea	Notes and further explanation
■ When there is no movement then all the forces acting are equal.	When objects are **stationary** it means that forces acting in one direction are **balanced** out by equal forces acting in the opposite direction. This results in their cancelling out each other, resulting in no movement.
■ **Resistance** is a generic term for any force which acts against movement.	When a force acts to **slow an object** down it is known as **resistance**. This can be air resistance or water resistance or electrical resistance; however, between solids it is more commonly known as **friction**.
■ Forces can be classified as those acting on **land**, or **water** or in the **air**.	**Land**: pushing, pulling, twisting, turning, upward, downward, backward, forward etc.
	Water: floating and sinking (sailing, steaming, propelling etc.)
	When an object floats then the force pushing it down (**gravity**), must be equalled by a force pushing it up. (**Upthrust** from the water underneath the object resisting the down force so that it results in a balanced force and stays on the surface). ▶

Big Idea	Notes and further explanation

Air: flying and air resistance. Flying is the same principle as floating, only it is in air and not in the water.

Gravity pulls the object down but the resistance produced by the air moving underneath the object provides **upthrust** and stops it falling. It can even overcome gravity if the wing design is right, and so the object will begin to ascend because upthrust is greater than gravity (10 Newtons).

■ The resulting movement of an object depends on the **size of the force** acting and how big and **heavy the object** is (Newton's 2nd law).

If an object is **heavy for its size**, i.e. dense, then it will sink because each bit of matter that makes up the object will attract a force of gravity. The more bits of matter in the same space the heavier it becomes. When it becomes heavier than water it sinks!

That is why a 50 g ball of plasticine will sink, but if we flatten it out and turn up the sides it will float. In the form of a ball it is 'heavy for its size', flattened out it is much 'lighter for its size'. (For 'size' read 'volume' to be scientifically correct.)

Big Idea	Notes and further explanation

It is the same principle if you want to stay up in the air. You have to move fast enough to create a big enough **upthrust** beneath your wings to overcome gravity's 'downthrust'.

Light and sound:

- Light travels from a source at an incredible speed.
- The Sun is the only source of the Earth's natural light and hence energy. (As mentioned in the Biology section.)

The **speed of light** is **186, 000 miles per second!** (Approximately 300,000 kilometres per second.) This means that if you could ride on a beam of light you would travel a million miles in just over 5 seconds! Even at this speed it takes the light from the Sun's surface about 8 minutes to reach the Earth's surface (96 million miles).

- The Sun is a star.

The Sun is just one of billions of stars in our galaxy, which is one of billions of other galaxies.

- The distance between stars in our galaxy and in the universe is so great that we measure it in **light years**.

A **light year** is not a unit of time, but a unit of **distance**. It is the distance travelled by a beam of light in one year. This incomprehensible distance is useful for measuring the distance of stars: because they are so far away the number would go on for ever if we did not standardise on a light year as the unit of measurement.

▶

Big Idea	Notes and further explanation
	The light that reaches us from other stars could have been travelling for thousands of years, so that when we receive it we are looking at what happened thousands of years ago!
■ The Sun is a massive nuclear reactor which radiates **electro-magnetic energy**.	Millions of nuclear reactions occur on the Sun's surface every second. The **radiation** from this process reaches us travelling at the speed of light. We receive the mainly harmless radiations of **light** and **heat** on the Earth's surface. The shorter wavelengths of X-rays and gamma rays would kill us if we were closer to the Sun!
■ Light travels in straight lines and in minute **waves** and can pass through some things but not others.	**Transparent** materials let all the light pass through them and are therefore see-through, e.g. glass. **Translucent** materials only let some light through and do not allow you to see through them clearly, e.g. bathroom windows. **Opaque** materials do not let any light through them and so we cannot see through them, e.g. bricks, metal, wood.

Big Idea	Notes and further explanation
■ The absence of light is darkness.	Since light travels in straight lines then if it comes up against an object which is opaque its energy becomes absorbed and nothing emerges at the other side, which appears in darkness, otherwise known as a **shadow**. (That is why black cars feel red hot in the sunshine!)
■ Light can be reflected off smooth shiny surfaces and produce a 'mirror image'	Light **reflects** off many different kinds of surfaces and bounces all over the place, giving the surface a shiny polished look. If the surface is extremely smooth then the light will bounce back all in the same direction and then we will see a clear image, albeit **reversed**.
■ White light can sometimes get split up into its colour components.	White light is made up of three different **primary colour wavelengths**, red, green and blue, which combine with each other in various degrees, to form all the other **colours of the rainbow**.
	When all three colour wavelengths are in equal amounts and they combine, the colour we call **white** is formed!
	When white light passes through some transparent objects some of the wavelengths are slowed down more than others: the white light gets 'fractured' and the different colours appear (rainbow effect).

▶

Big Idea	Notes and further explanation
■ Objects appear in different colours because the white light falling on them has had some of the colours absorbed.	The colours you see are the **wavelengths** that have escaped! If all the greens and blue wavelengths have been absorbed then you will see red!
	If all the colours have been absorbed then you will see black. Conversely, if all the colours are equally reflected you will see white.
■ Our eyes are made to collect the light that travels towards us allowing us to see what is in front.	Our eyes are very special cells that can respond to light waves, which the brain then converts into images for us. We call this process **seeing** and it has enabled us to survive for millions of years because we have been able to see danger coming and to recognise and avoid it.
	Logically this means that we are blind until the light passes through your pupil into the back of the eye.
■ Sound is another form of energy; it travels much more slowly than light.	Sound travels at over **300 metres per second** or over 750 miles per hour! Light travels at 300,000,000 metres per second, so sound is about a million times slower than the speed of light! (That is why we see the lightning long before we hear the thunder!)

Big Idea	Notes and further explanation
■ Sound energy causes molecules to **vibrate** and this vibration we hear as 'sound'.	Sound is also a form of wave, but unlike light it does not travel vertically but **horizontally**, just like a ripple when you drop a pebble in a pond. Again unlike light, sound **loses energy** as it travels and so gets fainter and fainter until it disappears.
■ Our ears are made to collect and capture the vibrations which our brain then interprets.	In the same way as for light, we cannot hear anything until it reaches our ears and the vibrations are passed on to the inside and begin to stimulate the brain. Once again, thanks to highly specialised 'hearing' cells we have managed to survive millions of years because we could hear and recognise danger, sometimes before it was close enough to see. Many animals have better hearing than sight for this reason.
■ To reach the ear, sound vibrations are passed along in a chain as it were, a bit like a 'Mexican wave'.	Sound travels by causing the molecules of the medium it is passing through to vibrate. Each molecule causes its neighbours to vibrate and so the excitation moves out as the pebble in the pond analogy described. ▶

Big Idea	Notes and further explanation
■ Some materials are better at passing the wave on than others.	Because the molecular structure of materials differ, some vibrate more easily than others and so carry sound better, e.g. sound travels better through water and steel than it does through air. This also explains why sound cannot travel through a **vacuum** where there are no molecules to vibrate. (Space is silent!)
■ Most of the time the vibrations are too small to see with the naked eye.	Normally these vibrations cannot be seen, with some exceptions such as stringed instruments; however, we can often feel them.
■ If we want to make the **volume** of the sound louder, then as we make the sound we need to give the vibrations more energy and make them bigger. This effect is known as **amplification**.	The louder the sound the bigger the wave it creates. Using the pebble in the pond analogy, if we gave it more energy by throwing in a concrete block we would get a much bigger wave travelling out from the centre and going further.
	On the other hand if we trap the soundwaves in a box with only a small hole to escape through then the vibrations go back and forth and multiply, producing a much louder sound, as in stringed instruments. Alternatively we can capture the sound electronically and 'amplify' it.

Big Idea	Notes and further explanation

■ If we want to alter the **pitch** from **low** notes to **higher** notes, then we have to make the vibrations quicker.

If the vibrations vibrate at a faster rate, e.g. not once per second but a hundred times a second, then the soundwave will travel at the same speed but with a much more energetic vibration, resulting in a higher pitch.

Lower notes are less energy demanding and hence will travel further before becoming exhausted. This explains why, when you hear the music from a passing 'boy racer' car, it is usually only the bass that you hear.

The Earth and beyond:

■ Our Sun is just one of the billions of stars in the universe.

We are part of our star's (Sun) **solar system**. Our Sun is one star in our **galaxy** in which there are millions of stars, and that galaxy is part of a **universe** which contains millions of similar galaxies!

Our nearest star is 4 light years away, but most of the stars you can see are hundreds of thousands of light years away, never mind those that are so far away we cannot see them! (Remember what a light year is? This will give you some idea of the size of the universe.)

The Sun is our **source of energy** and hence of life on Earth. If we were slightly closer to the Sun we would fry, and if we were slightly further away we would freeze.

▶

Big Idea	Notes and further explanation

■ Because the Sun is such a massive dense body it has a phenomenal gravitational pull and holds the Earth and the other **planets** in **orbit** around it.

The Sun is the **centre of our solar system**; this was a vital discovery by Galileo in the sixteenth century. Until then it was believed that the Earth was at the centre and everything revolved around the Earth.

■ The Earth's orbital movement is more or less the same each **revolution** and this predictability results in the **seasons**.

We move around the Sun at great speed but it still takes **one year** to complete a single revolution. This means that we are at a slightly different position relative to the Sun's surface at different times of the year and this results in our **seasons**.

At some times of the year we are a little closer and hence warmer, at other times a little further away and hence colder.

■ Not only do we go around the Sun at great speed, we also **spin on our own axis** at great speed and this produces day and night and causes shadows to lengthen and shorten.

Our planet Earth revolves on its own axis once in 24 hours. (Since the Earth is approximately 25,000 miles in circumference we must be spinning at over a thousand miles per hour! It's a good job the force of gravity is so strong or we would be thrown off!)

This means that for approximately half the **day** we will be facing the Sun and it will be **light** and warm; for the other half of the day we will be on the 'dark side' in the shadow, which we know as **night**.

Big Idea	Notes and further explanation

As we turn around and enter the daylight phase **the Sun appears low in the sky in the east** and produces **long shadows** in one direction. As we turn further the Sun **appears to be higher in the sky and overhead** and the **shadow** moves around and becomes almost **non-existent**. As we revolve further and look at the Sun from the opposite direction, **the Sun appears to be lower in the sky in the west** and the **shadows lengthen, but in the opposite direction** to the morning Sun.

■ In the same way as the Earth orbits the Sun, the Earth creates its own gravitational pull and this has resulted in the Moon being pulled into our orbit and revolving around us as we go along.

The **Moon** is our satellite in the same way as we are one of the Sun's satellite **planets**. The Moon makes **one revolution** around the Earth in approximately **28 days**. Because its position changes relative to us on the Earth's surface we see it differently at different times of the month, hence the **'phases of the Moon'**.

One side of the Moon is always **facing away** from Earth so we can never see this side unless we look from space.

▶

Big Idea	Notes and further explanation
■ Because of the 'spin nature' of all heavenly bodies, all stars and their planets are **elliptical** in shape.	As the Sun, Earth and Moon all spin very fast, they tend to form a circular type of shape; however, its diameter is slightly bigger in one direction than the other so they are not spherical but elliptical.
■ Distances within the solar system are immense; beyond the solar system they are unimaginable!	The Sun may appear to be close but it is almost 100 million miles away and even at the speed of light, light takes almost 8 minutes to reach the Earth's surface. Other stars are trillions of trillions of miles away which can only be measured in light years.

nb Throughout this section on Physics we have used the word **'energy'** on several occasions. This is an abstract concept which describes the motive force for all life. It can also be seen that this energy takes several forms and, more interestingly, one form can change into others and vice versa.

✳ eg Electricity can be converted to heat and light.

Mechanical energy can change into sound and heat, e.g. hammering.

Chemical energy can be converted into mechanical and heat energy, e.g. petrol to engine and wheel movement.

Heat energy can be turned into mechanical and then into electrical energy, as in power stations.

This concept of **energy** is a basic tenet not only of physics but also of biology (biochemistry and molecular biology) and chemistry (chemical reactions).

REMEMBER: SIMPLE SCIENCE CONCEPTS

QCA and DIY science

chapter 11

Don't be a QCA slave

The QCA Schemes of Work are non-statutory but nevertheless have proved to be very well used because they have provided a broad and balanced coverage of the science curriculum with clear progression from Year 1 to Year 6 and so also avoiding duplication.

The QCA recognises that 'one size does not fit all', and although advice is given on the DCSF Standards website (www.standards. dcsf.gov.uk) on how to adapt and customise its suggestions, it would still appear that a great many teachers have used almost word for word the contents of each unit. Unfortunately this can lead to a rather pedantic approach to teaching and learning, and because the teacher lacks ownership, their enthusiasm and creativity are diminished, and this in turn is reciprocated by the children.

The proposition of Section 4 is that although the QCA provides a sound planning structure, teachers need to inject their own Big Ideas and use more of their own investigations appropriate to their curriculum and their children.

It is hoped that with the newly found confidence from reading the previous sections, teachers will be encouraged to **Do It for Yourself!**

Keep it simple and look for the Big Science Ideas

The principles of planning with the QCA units are the same ones as we have applied throughout the book, i.e. KIS and look for the Big Ideas. This is similar to the seven-step methodology at the end of Chapter 1, particularly the first three steps.

Step 1: KIS and start with the Big Idea(s) about the science in the unit

It may very well be the same as the QCA suggestions or it may have a different emphasis or be less than or more than, depending on **your** needs.

nb Make sure that it is aligned with the appropriate part of the Programme of Study for that Year Group.

Step 2: Progression

You decide from the above, 'What I want my children to learn'. These are the key science ideas that are appropriate for this age group and sit comfortably with the topic, be it based on the QCA unit or part of a larger topic/theme based on the curriculum planning advice of the Rose Review (Rose 2009).

Step 3: Major concepts

In a similar vein you need to be clear about the key science concepts that are behind the Big Ideas and list them. There are two good reasons for this translation of Big Ideas into subject-specific concepts:

- It allows you to check the scientific **coherence** of the topic in a simple overview. Does it naturally 'hang together' and is the coverage in line with the school's science curriculum planning?

■ It states very clearly the scientific **outcomes** that you are looking for and focuses the scientific learning.

When you have decided the science aspect of the topic/theme informed by the school's plans for topics/themes and the appropriate QCA Science units, then we move to the next step which is to create links from the science to the wider curriculum.

Make links to the wider curriculum

There are two further steps in the planning process:

Step 4: Brainstorm other curriculum links

The cross-curriculum links can be based on subjects and/or curriculum themes as suggested by others, e.g. Rose (2009) and Alexander (2009), or a combination which suits the school. Look at each subject or theme and decide the other subjects' Big Ideas behind the learning. The learning is already prescribed in the various Programmes of Study and as has been said before, is enforced by law and therefore must be heeded.

In this brainstorming process the links to English and Maths have largely been ignored. This is for two reasons:

■ Teachers are more comfortable with these areas of the curriculum and so can make their own connections.

■ English and Maths are already heavily covered by the national strategies.

The same principle applies as in steps 1–3: identify the Big Ideas in the other curriculum areas then refine your brainstorm by removing those which are not appropriate at this time, based on coherence and progression. The results of this process will allow you to design **your own teaching unit** which will carry with it **your** enthusiasm and creativity.

Step 5. Design your teaching unit

This is an outline plan of the topic/theme based on a **matrix** formation where the **columns** represent the curriculum areas and the rows show the connecting links. Each **row** normally represents a week, but this need not be so, it will depend on the time available and the organisation of the teaching sessions.

At this stage the information is limited to a few word **'triggers'**, which will provide the stimulus for the more detailed lesson plans for each week.

When filling in the matrix it makes sense to put the subject that is driving the topic in the **central column** so that the balance and coherence is obvious.

This DIY approach has several advantages:

- It gives the planning focus and discipline which delivers the appropriate Programme of Study in a coherent and balanced manner.
- It provides a deductive approach to the planning, moving from the big picture to a more detailed subject perspective.
- It allows us to widen the curriculum coverage in a logical and meaningful way and makes more sense of the science involved.
- It provides a discovery approach more appropriate to the 'intelligent dialogue' pedagogy proposed earlier.

Apply an enquiry-based pedagogy

You cannot have a successful enquiry-based topic/theme unless you have an enquiry-based pedagogy! Common sense and logic tell us that if we plan a creative and stimulating topic but then deliver it in a pedantic way, we will not intellectually engage the children. The topic will then become moribund and children may regress to the copying out and colouring which was so heavily criticised by the report of the 'three wise men' (Alexander, Rose and Woodhead 1992).

The teaching strategies in Section 1 and the 'trios' at the start of the book will provide you with some of the flair and creativity not only to motivate children by inspired teaching, but also to achieve the desired learning outcomes and enhance your enquiry pedagogy.

In simplistic terms, try to put the 'thinking' first and the knowledge second.

REMEMBER: CREATIVITY; QCA; DIY

Examples of QCA/DIY topic planning

The QCA Schemes of Work are organised into units which cover the Science Programme of Study over six years. There are 38 units altogether. Each unit is broken down into sections; the sections provide the recommended teaching sequences and suggested activities. The Schemes of Work cover Biology, Chemistry and Physics in KS1 and KS2 and the units cover the relevant aspects of that subject. Within each unit there can be up to ten sections or teaching sessions to deliver that aspect of the subject.

The proposition is that if we take these very helpful unit suggestions and look for the simple Big Science Ideas, then we may be able to conflate and combine the unit with other curriculum areas and derive a plan more suitable for our children and more suited to the suggestions of the Rose Review (Rose 2009). Below are three examples to illustrate the **topic planning methodology**.

Unit 1A: Ourselves

Step 1: Big Science Idea

A brief overview of the ten sections in this unit provides us with an abundance of choice, e.g. alive/dead, growing, feeding, moving, basic anatomy, variety, adaptation etc.

The Big Idea would seem to be about **'living and growing'** and showing that we have similar functions to other animals; however, we all have different designs and ways of achieving this objective of **'staying alive'**.

We now need to consult our Programme of Study to calibrate the teaching sequence against these prescriptions. Once we have identified the appropriate possibilities we move to step 2 and decide the **possible** progression of the science, e.g.

KS1; Sc2: 1a → 1b → 2a → 2b → 2f → 4b

to name but a few!

Having **grounded** our possibilities in the QCA unit and aligned it with the Programme of Study we now have the confidence to design our own topic based on 'Living and growing'.

Step 2: Progression

Sc2; 1a: Alive/Dead

Sc2; 1b: Other animals that are alive

Sc2; 2a: External parts of humans and other animals

Sc2; 2b: Food and water

Sc2; 1b: Moving, feeding and growing

Sc2; 2g: Senses

Sc2; 4f: Reproduction

Sc2; 4b: Animals are grouped by their appearance

Step 3: Major science concepts and their Big Ideas

- **Life cycle:** all living humans and other animals grow, reproduce and die.
- **Life support:** all living humans and other animals need food, water and air.
- **Survival:** in order to stay alive humans and other animals need to move and use their senses to obtain food and water, find a mate and avoid being eaten themselves.
- **Reproduction:** all living humans and other animals need to produce offspring and ensure that they survive and grow or their families will die out.

Above are four scientific 'truths' which will provide a sound scientific basis for children learning about the biology of the human and animal survival process on Earth.

Step 4: Brainstorm other curriculum links

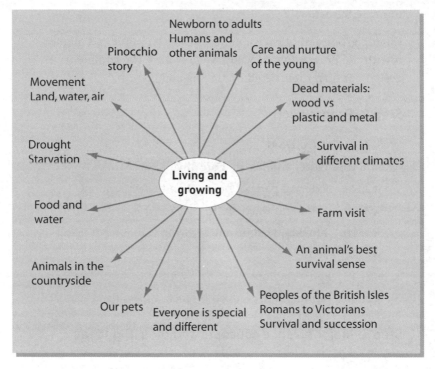

Figure 1: 'Living and growing' topic brainstorm of links to wider curriculum   Unit 1A

As said above, follow the steps 1–3 with other curriculum subjects and areas and produce a Brainstorm. Figure 1 is one possible example.

nb   You do not have to use all that you brainstorm. This is a creative free-thinking exercise which then needs to be honed to make it practical and fit for purpose.

Step 5: Design your teaching unit

Table 1 shows a topic outline which could operate over a half term in the afternoons. The timetable would determine the time available and hence the depth that you could go into. As science is the main focus this would need the most coverage; regarding the other suggestions, the teacher would decide whether to use a light touch or something more.

nb We have aligned the science with the appropriate NC PoS; however, we need to coordinate our plans with the Year Group and the Key Stage to make sure there is no duplication in the other curriculum areas.

Table 1: Curriculum balance matrix for 'Living and growing' Unit 1A

Miscellaneous	Human/social	Artistic	Science	History	Geography	D&T
Pinocchio			Alive and never lived	Peoples of the British Isles	The place where we live	
	Caring for each other: old and young		Food, water and air	Living conditions in the past from Romans to	Climate and food from other countries	Uses of wood, paper, metal, plastic
	Caring for pets					
	Disability and others less fortunate	Movement Dance	Moving: us and other animals	Victorians to today		
I am special						
	Safeguarding children		Growing: us and other animals			
Farm visit		Listening and music	Senses: us and other animals			Design a paper carrier bag
		Colour and painting				
Living, growing and caring			Life is a cycle			

End of topic plenary and displays

Unit 3F: Light and Shadows

This unit has ten sections which cover light travelling, the Sun, shadow formation, and reflection of light and seeing.

Step 1: Big Science Idea

The simple Big Science Idea is that **'light travels from a source at incredible speed and causes shadows'**.

We now need to consult our Programme of Study to calibrate the teaching sequence against these prescriptions. Once we have identified the appropriate possibilities we move to step 2 and decide the **possible** progression of the science, e.g.

KS2; Sc4: 3a → 3b → 3c

Having **grounded** our possibilities in the QCA unit and aligned it with the Programme of Study we now have the confidence to design our own topic based on 'Light and shadows'.

Step 2: Progression

Sc4; 3a: Speed of light; direct line of travel; Sun and manmade sources

Sc4; 3b: Light passes through some things but not others; Shadow formation

Sc4; 3c: When light comes to a shiny surface which it cannot pass through it bounces off

STEP 3: Major concepts and their Big Ideas

- Light is the Earth's only form of natural **energy** which comes from the Sun in the form of **light** and **heat**.
- There are other manmade sources of light.
- Light travels in straight lines at an incredible speed (186,000 miles per second, or 300,000 kilometres per second).
- The absence of light is darkness.
- When light cannot travel through an object (opaque), then

the area behind the object where the light cannot reach remains in darkness; this is called a shadow.

■ The size and direction of the shadow depends on the relative position of the direction of light to the object.

■ When the Sun is the source of light the shadow will change position according to where the Sun is in the sky.

■ Light can pass through some things to a greater or lesser degree (transparent; translucent).

■ If the opaque object has a shiny surface the light energy will not be absorbed but it will bounce off (be reflected) and be scattered.

■ If the shiny surface is smooth then the light will bounce off in the same order that it lands and the reflection will be a 'mirror image' which can be seen clearly.

Above are ten scientific 'truths' which will provide a sound scientific basis for children learning about the Physics of light.

Step 4: Brainstorm other curriculum links

As said above, follow the steps 1–3 with other curriculum subjects and areas and produce a Brainstorm. Figure 2 overleaf is one possible example.

nb You do not have to use all that you brainstorm. This is a creative free-thinking exercise which then needs to be honed to make it practical and fit for purpose.

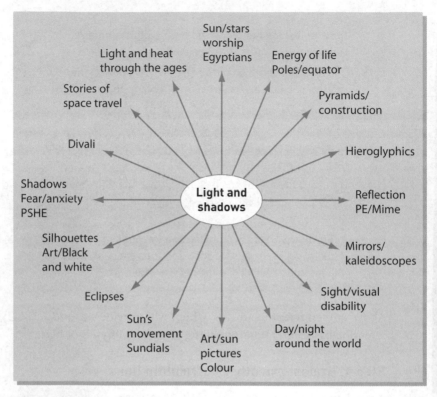

Figure 2: 'Light and shadows' topic brainstorm of links to wider curriculum   Unit 3F

Step 5: Design your teaching unit

Table 2 is a topic outline which could operate over a half term in the afternoons. The timetable would determine the time available and hence the depth that you could go into. As science is the main focus this would need the most coverage; regarding the other suggestions, the teacher would decide whether to use a light touch or something more.

nb   We have aligned the science with the appropriate NC PoS; however, we need to coordinate our plans with the Year Group and the Key Stage to make sure there is no duplication in the other curriculum areas.

Table 2: Curriculum balance matrix for 'Light and shadow'
Unit 3F

Miscellaneous	Human/ social	Artistic	Science	History	Geography	D&T
ICT: Egyptians research	Sun worship	Sun dance	Sunlight as Earth's source of light and energy	Egyptian culture and Sun Hieroglyphs Fire	Egypt on world map Equator and Poles Living in the hot	Pyramid structure
ICT: Famous artists		Artists and colour				
	Drought and starvation in the Third World	Light and shadow	Shadow formation investigations	Shadow clocks Time	Living in the cold Eclipses	Make a sundial
Scary stories	Divali Visual disability		Day/Night and apparent movement of Sun	Lighting through the ages		
Mirror writing		Mirror image dance mime	Shiny surfaces and mirror images			Making kaleido- scopes
End of topic plenary and displays						

Unit 6D: Reversible and Irreversible Changes

This unit has seven sections and deals with the basic chemistry of water and the concept of chemical reaction and change.

Step 1: Big Science Idea

There are two Big Science Ideas in this section:

1   Materials **react** with water in essentially one of two ways, they either dissolve or are insoluble, and if we want to **reverse** this then we have to use one of two basic analytical techniques, filtration or evaporation, to get our original materials back.

2   Some materials **react** with each other chemically and produce

changes which make them disappear or change into new materials and these chemical reactions may or may not be able to be reversed.

We now need to consult our Programme of Study to calibrate the teaching sequence against these prescriptions. Once we have identified the appropriate possibilities we move to step 2 and decide the **possible** progression of the science, e.g.

KS2; Sc3:   2d → 2f → 2g → 3a → 3b → 3c → 3d → 3e

Personally I would not deal with both these ideas in the same unit; the water chemistry and introduction of the concept of reversible reaction would sit more comfortably with evaporation, condensation and the water cycle in Unit 5D.

The second concept of 'chemical reaction and formation of new materials' is a basic tenet of Chemistry and I would focus on this aspect, particularly in the Autumn term around November!

Step 2: Progression

Sc3; 2f: Non-reversible changes

Sc3; 2g: Burning materials

Step 3: Major concepts and their Big Ideas

■ Materials may not react when mixed (inert); on the other hand, they can react with each other spontaneously and change.

■ Often if we wish inert materials to react and change then we need to add energy in the form of heat.

■ Many of these reactions are **irreversible** and the original materials change for ever into a new form and cannot be retrieved.

■ When sufficient heat energy is added to the material it will burn, which is the chemical process of combining with the oxygen in the air and is known as **combustion**.

■ If we contain the combustion process in a small space and stop the gases from expanding then it will eventually **explode**.

Above are five scientific 'truths' which will provide a sound scientific basis for children learning about the chemical reactions.

The progression of these Big Ideas will move through inert mixtures such as:

solids in water, e.g. tea and sugar;

liquid in water, e.g. orange concentrate and water;

solids with solids, e.g. pebbles and sand.

The next step will be to look at mixtures that spontaneously react without adding heat energy and change for ever, e.g. vinegar and bicarbonate of soda; plaster of paris and water; cement and water. (Often you can see and feel these reactions giving out heat energy.)

This would be followed by looking at some 'kitchen chemistry', the principle of which is to mix things together and **add heat energy** to get them to react and change irreversibly into something completely different that we can eat, e.g. Yorkshire pudding. (Man is always looking to make irreversible changes that produce new materials that are useful, e.g. polyethylene and nylon in the 1930s.)

Finally we will look at combustion, which is obviously an irreversible process, and see how useful it is to **release heat energy** to keep us warm and drive our machines. We will focus on fuels and link this to the by-products given off which can result in chemical pollution.

nb Use safety equipment at all times.

Step 4: Brainstorm other curriculum links

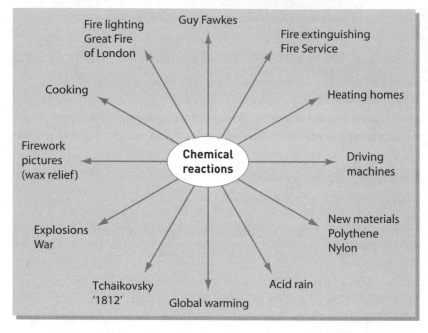

Figure 4: 'Chemical reactions' topic brainstorm of links to wider curriculum   Unit 6D

nb You do not have to use all that you brainstorm. This is a creative free-thinking exercise which then needs to be honed to make it practical and fit for purpose.

Step 5: Design your teaching unit

Table 3: Curriculum balance matrix for 'Chemical reactions'
Unit 6D

Miscellaneous	Human/ social	Artistic	Science	History	Geography	D&T
	Fire and safety	Mixing colours	Mixtures: Inert Spontaneous	Guy Fawkes		Building: Concrete Mortar
		Wax relief pictures (Fireworks)				
	Economic business: New materials		Kitchen chemistry (Mixing; dissolving; heating)		Acid rain	1930s: Polythene Nylon
					Global warming	
	Moral conflict	Tchaikovsky '1812'	Burning and releasing energy	Great Fire of London		Power generation
						Combustion engine
				Weapons of war		

Table 3 is a topic outline which could operate over a half term in the afternoons; however, the science would probably last less than half that time so it would seem to be sensible to use the science as a stimulus for the other work at the appropriate times. The timetable would determine the time available and hence the depth that you could go into. In the case of the other suggested curriculum links, the teacher would decide whether to use a light touch or something more.

nb We have aligned the science with the appropriate NC PoS; however, we need to coordinate our plans with the Year Group and the Key Stage to make sure there is no duplication in the other curriculum areas.

Section 4 has attempted to give you a simple methodology for designing your own themes and topics within the prescribed guidelines. If you wish to modify it in any way, please feel free.

Epilogue

In closing this book I should like to remind the reader of the aims that we started out with:

- To address and change the pedagogy needed if our children are going to enjoy and pursue science further and become independent thinkers and learners.
- To give teachers more confidence in their science knowledge and an understanding of the 'simple science ideas' at KS1 and KS2.
- To encourage and give teachers the planning skills to be creative with science and bring back enjoyment to them and their children.
- To show teachers that a basic philosophy of teaching children to reason logically permeates the curriculum and is the key to children's learning in the twenty-first century.

If I have achieved any of these aims to any small degree then I will feel that my book has not been in vain.

Thank you for reading it.

REMEMBER: IT IS ABOUT CHILDREN AND 'MAKING A DIFFERENCE'

Glossary of acronyms

APP	Assessment of Pupil Progress
CASE	Cognitive Acceleration through Science Education
DCSF	Department for Children, Schools and Families
DfES	Department for Education and Skills
DIY	Do It Yourself
ERA	Education Reform Act 1988
KIS	Keep It Simple
KS	Key Stage
KUSA	Knowledge, Understanding, Skills and Attitudes
NC	National Curriculum
OfSTED	Office for Standards in Education
PoS	Programmes of Study
QCA	Qualifications and Curriculum Authority
SAT	Standard Assessment Task
SSAT	Specialist Schools and Academies Trust

Bibliography

Alexander, R. (2009) *Towards a New Primary Curriculum* (Alexander Review), University of Cambridge Faculty of Education, available at www.primaryreview.org.uk.

Alexander, R., Rose, J. and Woodhead, C. (1992) *Curriculum Organisation and Classroom Practice in Primary Schools* ('The Three Wise Men Report'), Department of Education and Science.

Alexander, R.J. (2006) *Towards Dialogic Teaching: Rethinking Classroom Talk*, Dialogos.

Bullock, A. (1975) *Language for Life* (Bullock Report), Department for Education.

Clay, M.M. (1979) *Reading: The Patterning of Complex Behaviour*, Heinemann Educational Books.

de Bono, E. (1997) *De Bono's Thinking Course*, BBC Books.

DES (1988) *The National Curriculum Handbook for Primary Teachers in England, Key Stages 1 and 2, Programmes of Study*, Department of Education and Science.

DfEE (1999) *National Curriculum Handbook for Primary Teachers in England, Key Stages 1 and 2*, QCA/99/457, Department for Education and Employment.

DfES (2004) *Primary National Strategy. Excellence and Enjoyment: Learning and Teaching in the Primary Years. Learning to Learn: Progression in Key Aspects of Learning*, 0524-2004 G, Department for Education and Skills.

Donaldson, M. (1993) *Human Minds: An Exploration*, Penguin Books.

Driver, R. (1983) *The Pupil as Scientist*, Open University Press.

Feuerstein, R., Falik, L.H. and Rand, Y. (2004) *Creating and Enhancing Cognitive Modifiability: The Feuerstein Instrumental Enrichment Program*, ICELP Press.

Harlen, W. (ed.) (1985) *Primary Science: Taking the Plunge*, Heinemann Educational Books.

Harlen, W., Darwin, A. and Murphy, M. (1977) *Match and Mismatch*, Oliver & Boyd.

Lipman, M. (1988) *Philosophy Goes to School*, Temple University Press.

OfSTED (2008) *Success in Science*, The Stationery Office.

Rose, J. (2009) *Independent Review of the Primary Curriculum* (Rose Review), DCFS.

Shayer, M. and Adey, P. (2006) Cognitive Acceleration through Science Education (CASE), *Journal of Research into Science Teaching* 27(3): 267–85.

Smith, P. (2003) *An Introduction to Formal Logic*, Cambridge University Press.

Tharp, R. and Gallimore, R. (1988) A theory of teaching as assisted performance. In Tharp and Gallimore (eds), *Rousing Minds to Life*, Cambridge University Press.

Vygotsky, L.S. (1978) *Mind in Society: The Development of Higher Psychological Processes*, Harvard University Press.

Wood, D. (1986) Aspects of teaching and learning. In M. Richards and P. Light (eds), *Children of Social Worlds*, Polity Press.

Index

evidence-based conclusions 84–5
evolution, principle of 120
Excellence and Enjoyment strategy
  48, 63
experimental alteration 94
experimental design 70–1

'fair testing' 70, 93–4
Feuerstein, Reuven 63
'flat-lining' 106
flexible thinking 76
floating, science of 9–17, 23–6
force, scientific concept of 23–6,
  138–42
frameworks for learning xx

Galileo 150
Gallimore, R. 31–2
geography 26
global warming 127
graphs 94, 96
gravity 10, 115–16, 138–42, 150–1
ground rules for the classroom 50
grounded learning 16, 22
guesses, intelligent 88
'guided discovery' approach to
  teaching 2, 7

Harlen, Wynne 89
hidden curriculum xxiii
hierarchies of ideas 110

ice, melting of 45
'If … then' reasoning 53–4, 57–8,
  71
impressions, teachers' 107
impulsivity 75
inclusive education 48
incompatibility, logical 59
independent thinkers 77, 177
independent variables 70–1, 94
inferences 51–3, 61–2, 65, 86, 88
  alternative 66
integration of science with the
  rest of the curriculum xx, 22,
  161–2, 166, 169, 174
intellectual engagement 34–6
intelligent behaviours xxiii, 74–7,
  81–5

internalisation of behaviour 76
interpretation
  of attainment targets 99
  of scientific findings 44, 71
investigation, scientific 86–7, 90
  See also scientific enquiry

judgements
  arriving at 85
  critical 65–6

key science concepts 159
key skills 63, 90
key teaching points 13–14, 18
know; do; understand methodology
  xxiv, 100
knowledge, understanding, skills
  and attitudes (KUSA) concept
  7

law, criminal and civil 62
learning behaviours 75
learning objectives 12–13
  hiding of 6, 18
learning pathways xx, 13–14, 18,
  24, 82
'left field' thinking 67
lesson planning 7, 17–18, 24, 82
levels of attainment 99
light 17–18, 110, 143–6, 167–71
  speed of 143, 146, 168
light years 143
Linnaeus, Carolus 118
Lipman, Matthew 63
logic xxii, 36, 51–9, 64–9, 75, 177
  in reverse 67
  in test questionnaires 69
logical; critical; creative ways of
  thinking xxii, 63, 68
look; do; think process xxi–xxii, 45

magnetism 137
Match and Mismatch project 80
mathematics, nature of 58
measurement units 92, 138
Mendeleev, D.I. 124
micro-organisms 121
misconceptions in science 32
misinformation 67